Andrew S. Fuller

The Propagation of Plants

Andrew S. Fuller

The Propagation of Plants

ISBN/EAN: 9783744688970

Printed in Europe, USA, Canada, Australia, Japan

Cover: Foto ©berggeist007 / pixelio.de

More available books at **www.hansebooks.com**

THE PROPAGATION OF PLANTS,

GIVING THE PRINCIPLES WHICH GOVERN THE

DEVELOPMENT AND GROWTH OF PLANTS, THEIR BOTANICAL AFFINITIES AND PECULIAR PROPERTIES;

ALSO,

Descriptions of the Process by which Varieties and Species are Crossed or Hybridized, and the many Different Methods by which Cultivated Plants may be Propagated and Multiplied.

BY

ANDREW S. FULLER,

AUTHOR OF "THE GRAPE CULTURIST," "THE SMALL FRUIT CULTURIST," "PRACTICAL FORESTRY," ETC.

ILLUSTRATED WITH NUMEROUS ENGRAVINGS.

NEW YORK:
O. JUDD CO., DAVID W. JUDD, PRES'T,
751 BROADWAY.
1887.

TO

J. R. TRUMPY,

FLUSHING, N. Y.,

Whose great skill in the Propagation of Rare Trees and Shrubs has contributed so much to the tasteful appearance of the Homes of our People,

THIS VOLUME IS DEDICATED,

BY HIS FRIEND,

THE AUTHOR.

CONTENTS.

Preface	vii.
CHAPTER I.	
Propagation of Plants	11
CHAPTER II.	
Movement and Reorganization of Cells	20
CHAPTER III.	
Origin and Kinds of Buds	28
CHAPTER IV.	
Roots and their Functions	36
CHAPTER V.	
Stems and their Appendages	51
CHAPTER VI.	
Flowers, Fruits and Seeds	66
CHAPTER VII.	
Circulation of Sap	85
CHAPTER VIII.	
Sex and Fertilization	100
CHAPTER IX.	
Influence of Pollen	117
CHAPTER X.	
General Principles and Methods	135
CHAPTER XI.	
Propagation by Cuttings	144
CHAPTER XII.	
Propagation by Cuttings of Immature Growth	154

Chapter XIII.
Propagating by Layers 170

Chapter XIV.
Propagation by Suckers and Divisions 176

Chapter XV.
Propagation by Root-Cuttings 180

Chapter XVI.
Propagation by Budding 189

Chapter XVII.
Propagation by Grafting 199

Chapter XVIII.
Selecting Stocks 222

Chapter XIX.
Select List of Plants 249

Chapter XX.
Herbs, Tubers and Bulbs 309

PREFACE.

When I was a young man a plant of that grand old Rose, the Chromatella, came into my possession. Desiring to propagate it, I visited a Scotch gardener in the neighborhood and asked him to show me how Roses were budded, and was informed that propagating Roses was a "trade secret." This somewhat surprised me, as "trade secrets" in gardening was a new idea to me, not having at that time been introduced to the mystic shrine of the craft. My reply to this statement was, "Well, sir, if I live to be as old as you are, I will not only know your trade secrets, but make them known to all who may desire such information." The present work is in part a fulfillment of the promise made so many years ago, and it has never been lost sight of.

While admitting that a more intimate acquaintance with horticulturists and horticultural literature has somewhat modified my earlier impressions as to the general secretiveness of the profession, especially as to the more intelligent members, yet there are many who are still very chary of giving information about the best modes of propagating the plants they cultivate, and in visiting their establishments we may see "No Admittance" over the door of the propagating house, probably placed there to more fully impress upon the minds of visitors the idea that the proprietor or his gardener is an important personage, and that behind this sign are trade secrets of vast importance and value. But it will be found not only with horticulturists, but among all learned men, that he who knows most is the most willing and ready to give information to whomsoever may ask it.

The fountains of true knowledge are inexhaustible, and in practical matters it is folly to suppose that to know just how an operation should be performed will enable all to do the work equally well. One man may know how an implement should be made, and yet not possess the skill requisite to make it. It is said that poets are born, not made, but this oft-repeated aphorism is no more true of poets than of the mechanic or gardener. Long experience may enable a man to become a moderately good mechanic or propagator of plants, but never a first-class workman in either calling unless he possesses an inborn latent talent for such work that becomes developed through practical experience. For this reason there is not the slightest danger of the upper rooms in the temple of Hortulanus ever becoming overcrowded as the result of making known all that is possible to discover in regard to the cultivation and propagation of plants.

The present volume is a summing up of a life of observation, study and experiment among plants, in the field, forest and garden, and while in a few instances I may not agree with some of our botanical authorities, still, to be true to myself and my convictions, I could not do otherwise than state what appeared to me to be facts. It has been my aim in this, as in my other works, not to mislead, but to prompt the inexperienced to think as well as act—to investigate, experiment and seek the truth wherever it is to be found, without regard to what I or other authors have said. It is a common failing among cultivators of plants to consider words equivalent to action, and theories as facts derived from actual experience. We are all far too ready to accept theories in regard to the habits and structure of plants, instead of appealing to the plants themselves for the truth. It is so much more convenient to *believe* than to investigate and *know* whether a declaration is true or false, that we have comparatively few thorough and careful investigators of the phenomena of plant life.

Horticulture as a science is as yet in its infancy, and while we know something of the botanical relationships of plants as exhibited in the floral organs and various other appendages and parts, there is yet much to be learned of their chemical and mechanical affinities. In the hybridizing and crossing of species and varieties, and, perhaps, the intermingling of genera, there is a wide field open for investigation and experiment, from which very valuable and important results may confidently be expected; and while I have devoted only a limited space to this subject, enough has probably been said to show the way in which operations should be performed.

The usual incentive to investigation is a desire to know, and a doubt often becomes a germ of knowledge and an aid to progress. In this way have emanated most of the greatest discoveries of all ages. When one has become sufficiently interested in a subject to inquire and investigate, he enters upon the true and only road to actual knowledge.

In endeavoring to explain some of the physiological laws and principles which govern the growth of plants, I have not placed implicit confidence in the statements of those who are usually considered eminent vegetable physiologists, for it appears to be a common failing with the authors of such works to state definitely that a thing is thus and so, with seldom or never an *if, but,* or other modifying word that would indicate it was possible for the author to be mistaken, hence the gravest errors—if we call them by no worse name—have been widely disseminated and credited as absolute facts. I have stated only what my own experience among plants has led me to believe to be facts, without presuming upon the bounds of infallibility. The modes of propagation may not in every instance be the best known to others, but they are the best known to the writer.

The author asks no fellow laborer to place implicit confidence in the explanations given of the principles involved in the growth or modes of propagation of plants, but merely requests that their value or worthlessness be determined by actual personal experience, leaving all preconceived theories out of the question.

There was a time in the author's life when such a work as this would have been of great assistance, and have saved him the loss of many valuable plants; and believing that there are at this day young men just entering the horticultural field who may be benefited and derive some little assistance from this volume, it is sent forth with the hope that it will not only be kindly received, but may serve to promote the propagation and cultivation of plants.

ANDREW S. FULLER.

Ridgewood, N. J., January, 1887.

PROPAGATION OF PLANTS.

CHAPTER I.

PROPAGATION OF PLANTS.

For all practical purposes, the single individual plant-cell may be considered as the unit of vegetable organism. These minute individual cells contain a vital principle called life, which may be defined as a force, possessing an aptitude to respond to a stimulus.

It is the same in animals as in plants so long as they remain in their embryonic stages, but soon in the former energy, a condition of consciousness is delevoped. The egg of an insect, bird or other oviparous animal is a complex living organism, but it cannot in truth be said to possess consciousness, or be aware of its own existence; still, it responds to the stimulus (heat), energy is developed, and consciousness follows in succeeding stages. Seeds of plants respond to the same stimulus, the young plantlet absorbing nutriment from surrounding elements; and while it may not be conscious of the act, it possesses an inherent vital principle which enables it to reject certain substances and select others for its own use.

But as man has not as yet been able to analyze this vital principle which we term life, or trace it—this mysterious force manifested in matter—to its source, horticulturists are compelled to confine themselves to seeking a knowledge of the best means of promoting the development and multiplication of the cells, which, in their aggregate form, compose the plants that contribute so much to the welfare and pleasure of all members of the animal kingdom, man included.

The plant-cell, however minute, is not a solid body, nor composed of a single element, but in structure is made up of several parts, and these are the result of a combination of several substances. Within the young cell we find a viscid liquid, which has received the name of *protoplasm* (meaning formative matter), and usually floating in this there are numerous granules, the nature of which has not been fully determined, but they are supposed to be a kind of cell-kernels or nuclei, that probably play an important part in the production of new cells. As the cell acquires age and enlarges, the protoplasm forms a gelatinous coating on the inside of the true cell membrane, or what may be considered the proper wall of the cell. This membraneous inner surface of the cell-wall is called the internal utricle by Mulder, and primordial utricle by Mohl. It is only visible in new and very young cells, and soon disappears; but, when present, may be detected under the action of a tincture of iodine, which turns it yellow. As the cells thicken the internal utricle, also the cell-kernels or granules, become incorporated with the cell-walls. The chemical composition of the complete or mature cell is made up of three elements, carbon, hydrogen and oxygen; but in this young and immature state the protoplasm contains nitrogen, and the nitrogenous substances are known under various names, such as gluten, albumen, and other well-known products of vegetables.

Commencing with the individual cell, we find them in a vast number of plants so minute that they are invisible to the unaided eye, and their forms can only be determined by the aid of the most powerful microscopes. But down here, at or near the unit of vegetable life, we find perfect plants that consist of only a single cell, and among the most familiar of them is the yeast plant (*Torula cerevisiæ*) A cluster of these one-celled plants, highly magnified, is shown in figure 1. There are different species of yeast plants, each of a different form, and all may be propagated, under proper conditions, as readily as plants of higher orders. The Bacteria are among the most minute and obscure race of one-celled organisms, and the interest in these is increased on account of their frequent association with many of the diseases of the higher order of plants and of animals. There are hundreds and thousands of species of these minute one-celled plants, and they assume a great variety of forms; some are simple round dots floating in liquid, others in chains of cells, while some species are ornamented in the most intricate geometric patterns, while others are oval, long, or spread out in a fan-shape, as shown in the one-celled alga (*Licmophora splendida*), figure 2.

Fig. 1.—YEAST CELLS.

Fig. 2.—ONE-CELLED ALGA.

While these minute one-celled plants play an important part in the development and continuation of plant-life, still their small size has, in a measure, prevented a very general acquaintance with their structure and properties, as well as making it somewhat difficult to trace the relations of such low organisms through a multiplicity of channels up to the higher forms of plants.

SIZE AND FORM OF CELLS.—The size and form of cells are infinitely varied, and, as only a comparatively small number of the known species of plants have been subjected to a careful microscopic examination, we can only gather an idea of the many forms from what we know of the few. Wood cells of different forms and sizes are shown in figure 3. The size of cells may be said to vary from $1/_{20}$ of an inch up to $1/_{2000}$ of an inch in diameter, and there are probably some that are even of less size. Not only do the cells in different kinds of plants vary greatly in size, but such variation is also found in different parts of the same plant. It is also known that the cells of a rapidly-growing, healthy plant are larger than those of one that is feeble and sickly. In form, cells vary from the simple globule of the yeast plant in every conceivable direction, frequently assuming intricate geometrical figures, the ellipsoidal apparently springing from the spherical; following upward we find the cube, prismatic, hexagonal, stellate, fusiform, and branching cells. The spherical, oval and elliptical are most common in fungi and herbaceous plants, the more complicated appearing in shrubs, trees and other plants belonging to what are termed the higher orders.

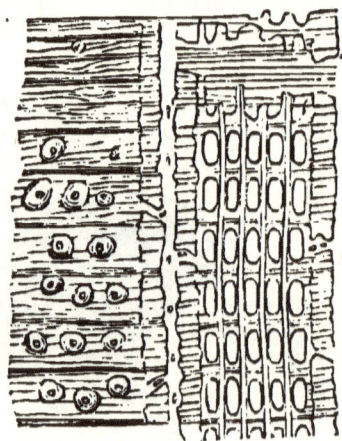

Fig. 3.—WOOD CELLS.

GROWTH OF CELLS.—Each individual cell is a direct

product of other cells, or some form of organizable matter containing the elements of which the cell is composed. At first the cell consists of a separate membrane, but as growth commences there is a re-arrangement of the formative matter, varying according to the characteristics of the plant in the course of development. In some of the lower forms, as has already been stated, the fully developed plant consists of a single cell, others of only an aggregation of the same or closely allied forms; but as we pass upward, the structure becomes more complicated, and there is a greater variety in the shape of the cells, as well as in the elements of which they are composed. The cells have the power of multiplication, new cells springing from the mother cell; these in turn producing others, and in plants like the common puff ball, many millions are produced in a few hours. Dr. Lindley calculated that in one gigantic species, the *Bovista Gigantea*, that the cells were produced at the rate of sixty-six millions in a minute. But in such simple kinds of plants as the mushrooms and sea-weeds, the entire structure is composed of what is called cellular tissues, a pulpy mass very similar to that which makes up the bulk of our cultivated fruits and vegetables. Some cells elongate to a great length and become a continuous hollow tube, as in the filaments of Cotton, or solid fibers, as in the inner bark of the Basswood and Papaw tree, or in such herbaceous plants as the Ramie, Hemp, Jute and other fiber-yielding species. There is a wide difference in the way cells are united. In some of the lower orders of plants the cell walls separate as they form, but in the higher and more complex, the walls of the young cells are solid and only divide or split apart as they advance in age. In many of the simple plants, the cells are widely separated and the intervals between them is filled with a semi-liquid mass, in which nothing that resembles a cell can be discovered. But as we advance to the higher orders of plants, a more compact and systematic organization is observable and the cell-walls touch each other at one or more points, permitting of the transmission of fluids from one to another. While young the cell-walls are so thin that they allow of a rapid transmission of fluids and gases, but when they reach a more mature condition the walls become thick and rigid, but never entirely impervious to liquids, for even dead plants will absorb moisture, and often assume the forms and colors which they possessed when alive. The restoration to apparent life of various species of Lycopodiums, Mosses, and the well-known "Rose of Jericho" (*Anastatica hierochuntinc*), are familiar examples of this kind. But it is only while the cells are young and contain protoplasm that they are available for the propagation of plants under artificial conditions. The propagator should keep this in mind, as it will often be of assistance to him in selecting cuttings and cions of plants for propagation.

THE TRANSUDATION OF FLUIDS.—The inherent power that cells possess of absorbing fluids and transmitting them from cell to cell, is the process by which nature enables plants to obtain nutriment from the medium by which they are surrounded, whether it be air, earth, water, or all of these combined. This transference of fluids from one cell to another, by a process of transudation, is universal among plants, and while it may be said that the energy displayed in the movement is controlled by a physical law applicable to both animal and vegetable membranes, still there is a vital force present in the living tissues, of the origin or properties of which we know but little. The operations of this force or principle may be, as is generally claimed by vegetable physiologists, purely mechanical, but that it does possess an inherent power of selection, using certain material or elements and rejecting others, can scarcely be doubted, or, as Dr. Thayer has said, in a paper on Plant Life, that,

"While some of the little cells are at work on delicious honey and rare perfumes, others are engaged in compounding healing medicines and even deleterious poisons." It is quite within the range of possibilities that plants possess the power of discrimination or choice, irrespective of their mechanical structure, and this may, in part at least, account for their responsive action to certain stimulants and not to others. Of course we are not to suppose that plants possess any functions corresponding with mind in animals, but they do possess a sensitiveness, which often nearly approaches, if it does not quite reach, the realms of intelligence. The physical process in obedience to which fluids pass from cell to cell, or through any permeable membrane, has been named *endosmosis* and *exosmosis*. The first is given to the inward flow and the other to the outward. These names were applied by H. J. Dutrochet, an eminent French physiologist, who wrote several valuable treatises on animal and vegetable physiology, published in Paris between 1824 and 1837, and since his time the above terms have been in common use. The explanation of this process of transudation is, that liquids of different density, placed on opposite sides of a permeable membrane, are naturally attracted or flow towards each other, and sooner or later become intermingled. The thinnest liquid will flow towards and into the thicker, and this movement is called *endosmose*, while at the same time a much smaller amount of the thicker will flow out into the thinner (*exosmose*), until both become of equal density. A linen or silk bag, filled with honey or thick syrup, and suspended in a pail of water will furnish a good illustration of the movements of fluids by transudation, for while the water will flow in, some of the honey will be dissolved, thinned, and then flow out, this process continuing until the entire liquid becomes merely honey-flavored or sweetened water, and of the same density throughout. The same kind of interchange, as it may be termed, takes place when the living tissues of any two closely allied plants are placed in contact, as in the common operations of budding and grafting, for it will seldom occur that the density of the fluids in both stock and cion, or even a bud, will be exactly the same; consequently, a movement, however feeble at first, must follow close contact of the living plant-cells. In some instances, as I shall have occasion to show in a succeeding chapter, the greater the difference in density of the fluids, the more likely are the two parts to unite quickly and permanently. But it is not to be supposed that the transmission of fluids from cell to cell is all that is necessary to make those of one plant support another, or to insure a union between the severed parts of two species or varieties, as in the operations of budding and grafting, for there is an individuality of plants not always easily recognized; still, it exists, and while the transudation of fluids may take place, there must also exist an affinity between stock and cion to insure the coalescing of cells. The principle involved cannot, with our present knowledge of life in plants, be fully explained, and vegetable physiologists usually refer to the movement of fluids in plants as a mechanical process, probably because this is the easiest way of bridging a chasm that they cannot fathom. The sap of the Oak may flow into a cion from a Hickory by the process of transudation, but the cells of the latter refuse to use it, or even respond to a stimulus from such a source. This individuality in the functions of cells enables one part of a plant to be engaged in accumulating very different elements from those of other parts. The leaves, flowers, roots and bark may all be manufacturing, as it were, quite different substances. The petals of the Rose emit a different scent from that of the leaf or other part of the plant, and that this fragrance is a distinct product of their cells is shown in the attar of Roses distilled therefrom. Then again almost every different variety of the Rose has an individual fragrance of its own, which is only emitted while the cells remain in their natural and original form. The bark-cells of many species of plants yield valuable products not to be found in any considerable amount in other parts of the same plant, as in the bark of the Cinnamon tree, Benzoin, Peruvian bark, and in the root-bark of the Sassafras. In other plants the roots may yield a large amount of coloring matter, as in those of the Madder, while only a trace of it appears in the stems and leaves. Seeds of some kinds of plants yield oil in the greatest profusion, with little or none in other parts; Flax, Rape and Cotton seed are familiar instances of this kind. Some seeds contain most powerful poisons, like those of the Strychnos, while the pulp surrounding them is innoxious. The Peach tree yields the most luscious of fruits, while from the seed may be extracted Prussic acid, the most virulent of all known vegetable poisons. Special deposits of special elements in the same plant may be considered the rule in nearly all of those most useful to man, and yet all these various substances are derived from the same sources, and are composed of simple elements, known by less than a half dozen different names. A plant, therefore, is in itself a chemical laboratory, and within its minute cells systematical evolutions are in progress, which we can neither see or fully understand, but the results are quite apparent to some one or all of our senses.

CHAPTER II.

MOVEMENT AND REORGANIZATION OF CELLS.

There is always a struggle in nature to right herself after any disturbing cause has interrupted her ordinary currents and conditions, and the propagator of plants is constantly doing this, especially when multiplying plants by division of cellular and woody tissues. Young and active cells contain matter that not only serves for the completion of the cell itself, but for the formation of new cells, and it is to the latter principle that we must attribute the ready response of vegetable structures to various stimulants and irritants. When a mass of cells is artificially divided or separated, as in the operation of making cuttings of the stems and roots of ligneous plants, or of cellular matter, as found in many bulbs and tubers, the exposed cells immediately make an effort to heal or cover the wound with new cells and restore the missing parts.

A severed root may, under favorable conditions, throw out new rootlets to take the place of the part removed, and from the exposed cells, made in removing a branch of a tree or shrub, new shoots often appear. Cuttings made of young twigs strive to furnish themselves with new roots wherewith to gather nutriment, and in all these various operations, there is a movement of cell-matter resulting in the production of new cells, no matter what form they may subsequently assume; whether it be that of roots, leaves, or any other of the many parts and appendages of plants.

The movement of cell-matter under multifarious conditions appears to be always in response to either a stimulant or irritant, and whatever other stimulant there may be present, that of heat is of paramount importance. It may be that only one or two degrees above the freezing point is all that is necessary to promote action in the vegetable cells of certain plants indigenous to cold latitudes, still, this is just as indispensable as sixty or seventy degrees higher temperature is for producing a similar movement in the cells of those inhabiting tropical climates.

It is to this movement of cell-matter, in response to a stimulant, that we are indebted for all the benefits and advantages derived from the artificial methods of propagating plants. The gardener's art consists principally in taking advantage of what he has here learned in regard to the natural functions and properties of plants.

The uniformity of the movement of cell-matter has enabled the propagator to formulate certain operations in order that they may be conducted under uniform conditions, and for the express purpose of producing uniform results. Under certain conditions, he is enabled to make the cells of one plant throw out new cells which unite firmly with those of another, thereby admitting of the passage of fluids, as in the operations known as budding and grafting. But under other and different conditions, this exuded cell-matter may become roots, capable of absorbing nutriment directly from surrounding elements in the soil, or, as in the case of the true epiphytes, from the atmosphere.

In the formation of knots on ligneous plants there is certainly a deviation from natural channels in the depositing of cell-matter. A few cells at first, through some obstruction or disturbing cause, are forced out of a direct course, and the next layer of woody tissues deposited must pass over or around the obstruction, this being repeated year after year, until the entire abnormal structure is complete and built up, in many instances, to an enormous size; for Ash-knots two to three feet in diameter, and weighing a hundred pounds, are not uncommon in many of the forests of our Northern States. The grain of these knots appears to be formed by cells propelled by some cyclonic force that gave to them a rotating motion, and yet there is a uniformity in this distorted re-arrangement. Another peculiarity of these knots is, that the bark covering them always partakes of the characteristics of the wood underneath and is fully as variable in structure.

But what may be termed the formative principle in plants, is always a potent power, and while frequently appearing to promote abnormal growths, is equally powerful in assisting the re-arrangement of cell-matter for continuing and perpetuating natural forms. The movement and re-arrangement of cell-matter under artificial conditions must have often been observed by all farmers and gardeners, although probably few of those engaged in these pursuits ever think these phenomena worthy of any special attention or study. An excellent illustration of this movement or action of cells is often found among the tubers of the common Potato, that have been stored in cellars or pits where the temperature was a little too high to insure perfect inactivity. An increase of temperature a few degrees above the passive point tends to incite action in the cells, and the most natural result would be a growth from the eyes or buds, which very frequently, if not generally, follows. But it is in the abnormal growths that we obtain the best illustration of the potency of the formative principle, as, for instance, in the not unfamiliar production of new tubers on the outside or within large hollow Potatoes. No leaves, stems, or roots are produced to assist the growth of these new tubers, but the starchy cell-matter of the old Potato moves inward or outward, as the case may be, and in reforming, builds up an entirely new structure out of the materials obtained from the old. These new tubers, produced in the absence of light, have well-formed buds or eyes on their surface, and, in fact, are in every respect typical representatives of their parents, and yet the conditions under which they have grown are certainly abnormal or unnatural.

Similar movements of cell-matter, as seen in the Potato, occur in other species of tubers and bulbs, and they may be looked for among all families of plants when placed under artificial conditions, or subjected to injury or serious disturbance of any kind. Propagators of plants can, and often do, utilize these abnormal growths, produced by the reorganization of cell-matter, in the multiplication of various species and varieties under cultivation.

Fig. 4.—BULB OF LILIUM SPECIOSUM.

For instance, with the scaly bulbs of Lilies, like those of *L. speciosum*, shown in figure 4, we have a complete and perfect structure which, if planted entire, will produce its one, or, at most, two flower stems, with leaves scattered along their entire length, the roots gathering nutriment from the soil for the support of the plant. But if we separate the individual scales and place them in a congenial material, such as moist earth, sand or moss, keeping them only moderately warm, and allow sufficient time for the change, each scale will produce a small bulb, an exact counterpart of the parent bulb. A small scale is shown in figure 5, as it appears when un-

Fig. 5. SCALE OF LILY BULB.

dergoing a change of structure. A cell near the base of the detached scale draws to itself the cell-matter of an adjoining one, the two add other cells, all deriving their support from the contents of the old cells, and from these reservoirs of nutriment a new scaly bulb of the exact type of the old or parent bulb is built up, just as occurred in the formation of a new tuber out of the materials of the old one in the Potato. Sometimes several bulbs will be produced from a single scale, as the formative principle may become active in one or more cells at the same time. These scales, in their aggregate form, make a complete or perfect bulb, are really sessile subterranean leaves, which, as shown, possess an inherent power of reproducing their kind in the absence of any additional organic material from other sources.

That the cell-matter of these subterranean scales or leaves is not specifically different from that of which the upper or stem leaves is composed, is readily shown when the old or main bulbs are planted so deep that a few inches of the base of the flower stem will be covered with soil. The flower stems of this species of Lily, as stated, produce leaves along their entire length, but those below the surface of the ground are prevented from developing into true aerial leaves; consequently the cell-matter provided by other parts of the plant is reorganized and becomes small bulbs (figure 6), which emit roots for their individual use and support later in the season, when they will be cast off by the mature and

Fig. 6.
LILY BULBS ON THE FLOWER STEM.

MOVEMENT AND REORGANIZATION OF CELLS. 25

ripened parent stem. That these stem bulbs are the product of reorganized cell-matter, which, under other conditions, would have spread out into long, thin, aerial and true leaves, is quite evident from the fact that they are only produced at the point where an embryo leaf-bud had formed a junction with the stem. It is not only in the propagation of plants under artificial conditions that we find great uniformity in the movement of cell-matter proceeding from uniform causes, but it is also observable in the results of attacks and injuries inflicted by insects. In the growth of what are called galls on plants, produced, so far as known, by the irritation caused in depositing eggs, or the presence of the larvæ hatched therefrom, the results are so uniform that the entomologist is enabled, at a glance, to identify the inhabitants of galls by the structure and outward appearance of their dwellings.

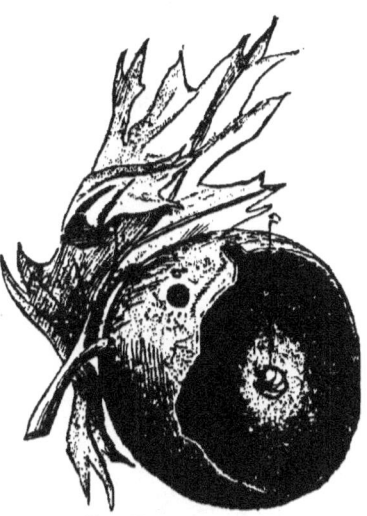

Fig. 7.—OAK GALL.

Why the irritation caused by the depositing of a few minute eggs by a small four-winged black fly (*Cynips spongifica*), on the leaf-stalk of the Black Oak should cause cell-matter to rush to that part, and form a large puffy gall an inch or more in diameter, and of a specific structure, differing widely from those produced by a closely-related insect upon other species of the Oak, we do not know; but the fact that every distinct species of Cynips produces a different gall is well known to every entomologist. (Figure 7—*a*, larva in center; *b*, hole where the fly escaped.)

In the true Oak Apple (figure 7), the cell-matter immediately surrounding the larvæ in the center of the gall becomes very hard and woody, while the space between this center kernel and the rind of the gall is filled with a grayish, light, spongy matter. In what is called the "Bastard Oak-Apple," produced by the *Cynips inanis* on the Red Oak, the central kernel or cell is not hard and woody, and the rind is connected with the center by slender radiating filaments, as shown in figure 8. Another species of Cynips produces a very distinct gall on the acorn of both the Black and Red Oak indiscriminately. Different species of insects not only produce different kinds of galls on the Oak, but upon various other kinds of trees and shrubs, and even upon many herbaceous plants. The Pine-cone Gall of the Willow is a familiar object in every swamp where the Heart-leaved Willow (*Salix cordata*), is found, and is the result of the depositing of eggs of a minute species of gall-gnat. As soon as the eggs are deposited in the young twig of the Willow, the cell-matter proceeds to form a cone-like structure, something entirely different and entirely foreign to the natural functions of the plant, and this movement continues until the gall is complete, the outer part being covered with regularly formed scales, the whole resembling a small pine cone. There may be thousands of these galls in the same field or swamp, but all will be found nearly of one size, and built up on the same general plan.

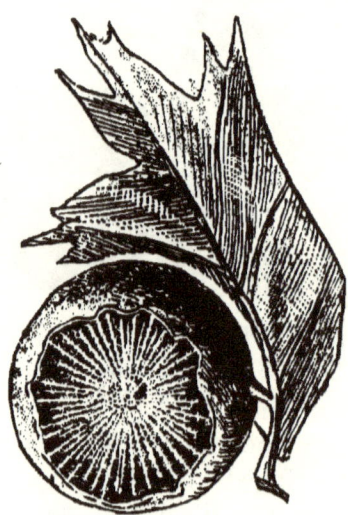

Fig. 8.—BASTARD OAK APPLE.

Where different species of insects, so far as our knowledge enables us to determine, produce galls out of identically the same materials, they are invariably of a different structure and form, showing that however slight the variation in the chemical or mechanical nature of the cause, the results may be widely variable. For instance, there are several species of gall-gnats that breed in the Grape-vine, but each produces a distinct form of gall. One species that attacks the young succulent canes makes a gall somewhat resembling a small apple, and it is known as the "Apple Gall" of the vine, while the habitation of another species resembles a cluster of fil-

Fig. 9.—TRUMPET GALL ON GRAPE-LEAF.

berts. The same rule holds good with galls produced on the leaves of the Grape-vine. The galls produced by the minute parasite formerly known as *Pemphigus vitifolia*, Fitch, but now as *Phylloxera vastatrix*, Planchon, appear on the underside of the leaf, and are merely small green fleshy swellings, more or less wrinkled, and with a slight depression on the upper side, forming a cup with a kind of hairy or pubescent margin. But in another leaf-gall, known as the "Trumpet Grape-gall," produced by a small species of Gall-gnat (*Cecidomyia*), the eggs are deposited on or in the upper surface of the leaf, and the cell-matter, moving to the point of irritation, builds up a trumpet-like gall (figure 9), a quarter to one-third of an inch in length. These galls, when produced on the leaves of wild vines having reddish petioles or leaf stalks, assume the same color, and on others they are greenish white, showing that the cell-matter, when responding to an irritant, may not change its natural chemical properties any more than it would in responding to a stimulant, and guided by the formative principle while perfecting the growth of the plant.

We are not to suppose that these galls are necessarily injurious to plants, for while the normal conditions and channels of circulation may be temporarily changed, followed by a re-forming of cell-matter under new conditions, and for the express purpose of giving food and protection to an insect, still the tissues of which these galls are composed are generally healthy, and are, without doubt, capable of performing the regular functions of assimilation.

CHAPTER III.

ORIGIN AND KINDS OF BUDS.

The origin of buds is a subject that has received much attention from vegetable physiologists, but in pursuing their investigations it is to be feared that many of them have been more anxious to confirm some previously conceived theory than to discover the truth from actual personal research.

Beginning at the unit of vegetable structures, we find plants composed of a single cell. In the multiplication or growth of these single cells a new cell is produced—

in other words, it puts forth a bud, this in turn another; these, however, are only single cell buds, but as we advance upward in the scale we find plants composed of an aggregation of cells, therefore compound in their structure, and their buds, assuming the parent form, are also made up of a number of cells, arranged in regular order, through the controlling influence of what is called the formative or living principle. Keeping the fact in mind that new cells are always the product of other, or parent cells, and that in the re-arrangement of cell-matter, or growth, the new cells preserve their typical form, we can readily understand how buds, even of the most complicated structure, may, under favorable conditions, be produced from living organizable vegetable matter, on or within any part or appendage of a plant. It is true that, when plants are under normal conditons and with room for full development, they produce buds uniformly at certain points and not elsewhere; but it is seldom that a plant is so favorably situated as not to be disturbed at some period in its life; consequently, abnormal growths may become hereditary through the oft-recurring influence of abnormal conditions. In a state of nature, there is usually more or less crowding among plants, and parts are disturbed, broken, or destroyed. The larger animals trample upon the small and young plants, or later in life browse upon the stems and branches, while insect enemies sting, cut and wound plants in various ways, consequently there is always a struggle for existence, which begets a necessity for a departure from what, under other conditions, would be the natural order of growth.

There are as many different forms of buds as there are different species of plants, but for the sake of convenience, they may all be arranged under five groups, viz.: The (1) terminal, (2) axillary, (3) accessory, (4) fruit bud, (5) adventitious. Buds, although of various forms, are generally, in the higher order of plants, composed of imbricated scales, enclosing the rudiments of stems, leaves, or organs of reproduction. The terminal bud is situated on the end of a stem or branch (as in fig. 10), and the axillary bud on the side just below, and usually in the axils of the leaves, and this bud may either grow and become a branch or remain dormant, or be entirely overgrown and obliterated. The terminal and the axillary buds are really the same in structure and importance, for if the terminal bud be destroyed, the one next below it becomes the terminal, and the elongation of the stem or branch proceeds with but a slight check. It is seldom that all of the axillary buds become branches, as only the strongest and most favorably situated grow, for nature, in her prodigality, provides more than are required in the regular order of growth. The accessory bud is merely one of a cluster of buds sometimes surrounding the base of a terminal, but more frequently accompanying the axillary buds. If by accident or otherwise the regular terminal or axillary bud is destroyed, the accessory buds take their place, one or more of their number growing. These accessory buds are not only found on the stems of annual and perennial plants, both herbaceous and ligneous, but also scattered over the surface of such tubers as the common Potato and Jerusalem Artichoke. In other tubers and bulbs they are found clustered at the crown or apex, as in the tubers of the Dahlia, Herbaceous Pæony and bulbs like the Turnip, Beet, Crocus and Gladiolus. Then, again, we find the axillary and accessory buds distributed along subterranean stems of various kinds of plants, and naturally they are only produced at the nodes or joints, as seen on the cane of the Grape-vine

Fig. 10.
TERMINAL AND AXILLARY BUDS.

ORIGIN AND KINDS OF BUDS.

Fig. 11.—WOOD SORREL (*Oxalis Acetoxella*).

or the underground branches of the Canada Thistle and common Wood Sorrel (*Oxalis Acetosella*), shown at *a* and *b*, figure 11.

The propagators make extensive use of all these forms of buds, in multiplying the various kinds of plants under cultivation.

Fruit-buds appear on plants in the same position as other buds of the regular form—that is, they are either terminal or lateral, but principally the latter; on trees they are often found on short spurs, which remain productive for many years. These buds contain the embryo organs of the flower, and in many kinds of plants they are formed the season previous to their full development. But the production of visible fruit-buds is by no means a universal characteristic of plants, for in many orders, and especially among cryptogamous plants (Ferns, Selaginellas, Mosses, etc.), true flowers or flower-buds are unknown, for the spores, which answer the purpose of seeds in the higher orders, are developed on the stems, leaves or other appendages of the plants, without the appearance of any previously formed organ in the least resembling flower or fruit-buds. Still, some of these plants produce both aerial and subterranean buds as freely and with as great uniformity as the most common of our cultivated fruit and ornamental trees and shrubs. The spores of these cryptogams are also produced with as great regularity and distinctness of form and position on the plant, and are usually as available for the purpose of propagation, as the most perfect and highly developed seeds.

Adventitious buds have long been a source of discord among vegetable physiologists, and some of the old and erroneous theories are not as yet quite obsolete. It is less than a century ago that several learned European botanists claimed that every bud on the stem of a tree was an embryo plant fixed in position, but sending its

ORIGIN AND KINDS OF BUDS. 33

roots downward and its leaves upward, having a kind of individual existence in the performance of its natural functions. All buds that might appear on the stem, and of the regular order, were supposed to be developed dormant or latent buds that had remained inactive from the time they were originally produced on the young stem when it was first clothed with leaves. The idea that buds could be developed or formed out of the young cell-matter at any time or at any point, on certain kinds of plants, seems never to have occurred to many of these earlier investigators, and, I may add, even to some still living. That buds do lie dormant for a time—years in some instances—must be admitted, but this will not account for their frequent appearance on parts of plants where no bud could have previously existed, as on the internodes of the stems and branches, as well as on leaves and true roots. In seeking for the origin of these adventitious or chance buds, we have no occasion to confine our investigations to a single order of plants, because they constantly appear in a vast number of species distributed

Fig. 12.—ROSE-LEAF AS CUTTING.

among widely separated genera and families, and it is not at all difficult to trace their source to the reorganization of cell-matter. The leaf of a Rose cannot be said to contain buds of any kind, still it will, when stripped from the stem, with no wood or bark attached, and placed under favorable conditions, produce roots and a bud near the base of the stalk or petiole, as shown in figure 12. The same is true of the Bryophyllum, some Begonias, and hundreds of other familiar and well-known plants that florists are constantly propagating by leaves, and even minute portions of a leaf. Even the bark taken from the internodes of some kind of plants, where there are no indications of the existence of dormant buds, will produce both roots and buds. Were it not for our modern facilities—notably, the microscope—we might be inclined to account for the origin of these chance buds by supposing, as claimed by some of the earlier botanists, that the germs or embryo buds were floating about in the sap of the plants, ever ready to develop when an opportunity offered.

H. F. Link, a celebrated German botanist of a half century ago, says: "The adventitious buds are distinguished from the axillary buds by their structure; in the latter, the greater part of the pith goes with the wood into the supporting leaf, and in the former the entire amount of pith passes into the bud." But Prof. Link's distinction between axillary and adventitious buds were more imaginary than real, and that the pith of even ligneous plants supporting the leaf was long since shown to be erroneous. Dr. J. M. Schleiden, in his "Principles of Botany," published about forty years ago, in speaking of adventitious buds of perennial plants with vegetation periodically dormant, says: "that the axillary and adventitious buds are only to be distinguished by the mode of development," and further that, "each stem, whether a common one or a root stem, can develop a bud. These buds are caused, not only by accidental and intentional wounding of the stem, but also by the inclination of plants to develop buds at certain places." However much later botanists may disagree with Prof. Schleiden in regard to his pet theories and caustic denunciation of some of those advanced by his contemporaries, practical horticulturists of the present day will certainly agree with him as to the appearance and origin of adventitious buds, although, in the chapter previous to the one from which the above remarks are copied, he states distinctly that, in Dicotyledons, "no root is capable of producing buds," p. 220. But, as I have said, adventitious buds appear as freely on roots as on leaves, and under similar conditions, but vegetable physiologists seem to be inclined to ignore the fact, or get around it by saying that when a root produces a bud it should be considered as a subterranean stem. The late Dr. Lindley in his "Theory of Horticulture" says: "In general, roots have no buds, and are therefore incapable of multiplying the plant to which they belong," but he adds, "that it constantly happens in some species that they have the power of forming what are called adventitious buds; and, in such cases, they may be employed for the purposes of propagation."

Prof. Asa Gray, in his "Structural Botany," Chap. iii., p. 82, says that, although roots are not naturally furnished with buds, yet, under certain circumstances, those of many trees and shrubs and some herbs, have the power of producing them. Again, in Chap. iv., p. 98, he says of adventitious buds: "It has been already remarked that roots, although naturally destitute of buds, do yet produce them in certain plants, especially when wounded." This is very true, but there are many species of plants, and notably among these the Wild Raspberry of both Europe and America, that multiply naturally and rapidly from adventitious buds on their true roots, as well as from those developed on what are usually considered as subterranean stems.

With the present state of our knowledge in regard to the structure of plants, it would not be wise, to say the least, to attempt to fix a limit to the range of adventitious buds, for, as I have already said, their origin depends largely upon the vital energies of the parent cell, and, as Prof. Schleiden remarks, "the self-subsistence and power of reproduction of the cell is the foundation of the reproduction of plants." From this power, under favorable conditions, can each individual cell or group form new cells and buds, which may or may not, according to circumstances, become a new plant. Botanists have been inclined to pass over the adventitious bud as a thing rarely occurring, and then merely a chance production, consequently not of sufficient moment to call for any special attention, but to the propagator of plants it has become of great importance, wherever and whenever it can be made available in the multiplication of valuable species and varieties.

CHAPTER IV.

ROOTS AND THEIR FUNCTIONS.

In many of the simple plants no organs corresponding to the true roots of the higher orders are produced, the nutrient elements being absorbed directly through all parts of the plant alike. At what point in the ascending scale the true roots first appear has not been fully determined, but these organs most probably follow, or are coexistent with, the inception of a division of cells into groups, each possessing special functions, as found in all compound plants.

As soon as a plant has advanced upward in the scale, and reached a plane where different organs are evolved with distinct and special functions, roots become necessary to its existence, and are consequently among the first of a multiplicity of organs to appear. It is true, if we consult the writings of our most eminent vegetable physiologists, we find that they do not agree as to what part or how much of certain plants should be designated as roots. In the funguses, such as mold and mildew, as well as in the larger mushrooms, the filmy threads called *mycelium* answer the purpose of true roots, producing a form of bud from which spring up, in some instances, stout stems, bearing a cap containing fruit-like organs; while in others there is only a very minute ascending axis.

While authors differ very widely in their opinions as to what part of certain plants should be considered as roots, and what should not, nearly all agree that whatever is developed below the cotyledons, figure 13, or first pair of seed-leaves, in dicotyledonous plants, should at first be considered as the root or descending axis; but this part soon undergoes a change, or outgrows its normal functions, and becomes merely an extension of the true stem or ascending axis. The embryo root, no doubt, when it first pushes out from the seed, absorbs nutriment from the elements with which it is surrounded, just as the smaller rootlets do later in the life of the plant; but when this main or central root ceases to absorb nutrients it would be difficult to determine. The stem or ascending axis, and the two cotyledons (seed-leaves), with a central terminal bud or leaf (plumule) is that part of the plant which seeks the light and air; while the part growing in the opposite direction, or downward, is the root or descending axis or radix. In many of the monocotyledonous plants (one cotyledon or seed-leaf), like the Grasses, Asparagus, Yuccas, Palms, etc., there are a number of roots, or a multiplication or division of the radix from the first, and, in some instances this emission of what may be properly called adventitious roots appears to be the normal habit.

Fig. 13. — APPLE SEEDLING, SHOWING COTYLEDONS.

But these rather anomalous characters need not surprise the propagator of plants if he will keep in mind that variation is a rule of nature; and while there is sufficient uniformity in the laws governing the growth of vegetation to enable us to discover what we may term general principles, yet it is impossible for us to determine with certainty the exact limits of variation.

In a general way, it may be said that roots growing in the ground have a tendency to go downward, or toward the center of the earth. Prof. Gray says that the plantlet possesses a kind of polarity, and is composed of counterpart systems—namely, a descending axis, or root, and an ascending axis, or stem. Prof. Balfour, in referring to this subject, says: "Physiologists have not been able to detect any law to which they can refer the phenomena, although certain agencies are obviously concerned in the effect. Some have said that the root is especially influenced by the attraction of the earth, while the stem is influenced by light." But experiments have shown that the downward course of the root is not always due to the attraction of gravitation, or to moisture in the soil; neither is the ascent of the stem due to the action of light, although all these, no doubt, have an influence upon the plantlet in its early stages of growth. Thomas Andrew Knight placed mustard seeds and French beans on the circumference of two wheels which were put in rapid motion, the one horizontal and the other in a vertical manner, and he found that in the former the roots took a direction intermediate between that impressed by gravitation and by the centrifugal force—namely, downward and outward, while the stems were inclined upward and inward. In the latter the force of gravitation was neutralized by the constant change of position; the centrifugal force acted alone, by which the roots were directed outward at the same time that the stems grew inward. But these variations may have all been due to the liquid condition of the parts of the young plants. Some plants grow indifferently in all directions from the very inception of germination, and it is well known that the roots of parasites point toward the center of their host. While the first root of the plantlet growing in the earth, and in a position free to act, generally goes downward, the secondary or adventitious roots have a tendency to wander in search of food, not only pushing out at right angles from the central descending axis, but often climb up banks at a very acute angle. There is no doubt about the uniformity of action in first roots of plants, but those produced later are controlled by varying conditions and circumstances.

The sensitiveness of the root is mainly in its point, and it is through the younger parts that the nutrient properties are mostly absorbed from the medium in which they live. Old or large roots cannot take up moisture or the elements of nutrition necessary for the growth of the stem and its appendages; hence the necessity of preserving the small fibers from injury in transplanting, as well as placing the principal roots in a position where new rootlets will be produced in time to respond to a call upon them for nutriment by the leaves.

The most natural position for roots is at the base of the stem, whatever form the latter may assume, whether it be that of a tree, shrub, vine, herb, bulb, tuber, epiphyte, or even a parasite, which pierces with its roots the tissues of its host for nutriment. But as all roots are produced by the multiplication of cells within restricted limits, they may, like the adventitious bud, appear, under favorable conditions, upon all parts of the same plant— on the stem, leaves, buds, or other appendages. Roots have no more fixedness of character than branches, for they may in many instances be changed into stems or branches, and there are some kinds of trees, like the Willows and Poplars, that may be completely inverted, the roots becoming branches and the branches roots. The same is true of many other kinds of plants to which I shall have occasion to refer hereafter.

Roots, like stems and branches, lengthen at their points only, and while the absorption of moisture and nutriment is principally through the newly-formed cells near the ends of the rootlets, still they continue to take in liquids through all parts until they have become hardened and enclosed in a cortical layer of cells. It is no doubt true, as Mr. Darwin claims, that sensitiveness to moisture resides specially in the tip of the root, but it can readily be shown that the absorbing property is not wholly confined to this point, by removing the tip of the rootlet. The absorbing powers of the young fibers must be far greater than could possibly exist in the extreme point, in order to supply the loss of moisture through rapid evaporation from the leaves and young twigs of many kinds of herbs and trees during dry and hot weather. But at what age or period of growth the rootlet loses its power of absorbing nutrients has not been determined, but it is probably variable in different kinds of plants. The epiphytes, and especially the larger species of Orchids, with coarse, fleshy, aerial roots, retain the power of absorbing liquids through their side cells for a much longer period than those of plants growing in the earth, having roots so minute that they can readily pass between small particles of soil while seeking sustenance. They not only add new cells to their points, but the cells in the rear are continually throwing off new branches, thereby enabling the plant to occupy and gather food from new sources.

The cause of this rapid formation and multiplication of absorbing points on many of the most vigorous perennial plants is unknown, for they are evidently wholly adventitious, not originating from buds like the natural branches on the stems of plants; and, furthermore, the larger proportion of these minute roots are deciduous, and only serve a temporary purpose, living but for a year or two; a few of the stronger remaining permanently, the others dropping off when no longer of service to the plant. The number and ramifications of roots are naturally quite variable in different kinds of plants, as well as in the same species growing under different conditions, and the cultivator frequently takes advantage of these variations in many of the operations which he is called upon to perform.

The principal office of roots is to collect nutrients from the medium by which they are surrounded, furnishing a vehicle for conveying the materials collected to other parts of the plant. But in many instances the roots also act as a support and for securing the plant in a permanent place or position. This function of localization, however, is not general throughout the vegetable kingdom, but belongs to certain groups or families, and even with these it is only operative during the life of the individual plant. Collectively, all plants may be said to travel or change places with each successive generation, and there is a vast number which are not confined to one spot for any considerable period during their lives, but they are continually moving from place to place, as seen in many aquatic plants floating in ponds, rivers and bays in all parts of the world. There are also many parasitic plants, like the common Dodders (*Cuscuta*), which at first spring up from seed buried in the earth; but the plant soon breaks loose from the parent root, then moving onward over its host, from which it obtains nutriment, the older parts of the stem dying and dropping off as the younger parts advance. Such plants may be said to travel in search of victims, leaving only their seeds scattered along the way, while other plants, like the common Black-cap Raspberry and Trailing Blackberry, leave one of their kind at each place occupied. The old plant sends out a long slender shoot, several feet or yards in length, and from the very tip of this, new roots are emitted, and in this way a new plant is established far away from the parent stock. While such plants do not migrate, or move about in the same manner as the floating aquatic herbs, still each successive generation seeks a new abiding place, at a greater or less distance from the homes of their immediate progenitors.

Wherever a plant becomes established, the roots gather nutriment from the medium in which they are placed, and of such nature as is required to build up the structure. These nutrients are absorbed or taken in through the surface of the roots, and as all must pass through the minute cells, it is quite evident that they can only be utilized when in a liquid or gaseous condition; nothing of a solid nature can be appropriated for use by the plant. Everything found in a mature plant must have originally entered the root or other parts as a liquid or gas, and then changed, by some chemical or other agency, into whatever form it afterward assumes.

Whether roots possess an inherent power of selecting their food, or not, is still a mooted question among vegetable physiologists. Some are quite positive that they do possess this power in a greater or less degree, while others are just as certain that they do not, and that all matter presented to them in a liquid or gaseous form is alike absorbed; hence the frequent cause of death among plants through the absorption of poisons. Dr. W. B. Carpenter, in his "Vegetable Physiology and Botany," in referring to this subject, says: "that they appear to have a certain power of selection; some of the substances dissolved in the fluids which surround the roots being absorbed and others rejected. Thus, if a grain of wheat and a pea be grown in the same soil, the former will obtain for itself all the silex or flinty matter which the water of the soil can dissolve; and it is the deposition of this in the stem which gives to all the grasses so much firmness. On the other hand, the pea will reject this, and will take up whatever calcareous substances (lime and its compounds) the water of the soil contains."

Prof. J. H. Balfour, in his "Manual of Botany," says: "Gaseous matters are taken up by the roots of plants and circulated along with the sap, as well as in the spiral vessels. These usually consist of common air, carbonic acid, and oxygen." And further, he thinks that the differences in the absorption of solutions depend on the "relative densities alone, and not on any peculiar extracting power of the roots, for it is well known that poisonous matters are absorbed as well as those that are wholesome." On the contrary, another English authority, Dr. Maxwell T. Masters, in his recent work, "Plant Life," says: "It is a moot point whether any carbon is taken up by the roots, but if any, it is only a small proportion." But Dr. J. M. Schleiden ("Principles of Scientific Botany"), in referring to this same mooted point, says: "The most universally distributed medium of solution in nature—water—is also the fluid which is absorbed by the plant cell, and conveys all other matters into its interior. The most essential of these matters are carbonic acid and ammonia, both of which are contained in water which either falls from the air or has been a long time in contact with it. Water, carbonic acid and ammonia contain carbon, hydrogen, oxygen and nitrogen, all of which are essential to the formation of the assimilated substances, and to the especial nourishment of the cell. But water occasionally conveys to the cell, in small quantities, all substances which are capable of solution in water."

Of the many other works of equally celebrated authorities examined on this point, no two fully agree in their "opinions," for we can scarcely bestow upon the information derived from such sources so dignified a name as "knowledge." But we may safely credit roots with the faculty of modifying and changing certain elements in their passage through the cells, and the chemical, mechanical and vital forces are all engaged in this work. There is a thorough filtering of the solutions as absorbed, else the crude sap or liquid would often remain colored as it passes upward; but this seldom occurs, even when the roots are submerged in highly colored, and what are generally considered, very nutritious fluids, as, for instance, the drainings of a manure heap, and from other similar vegetable matter. Still, roots do sometimes absorb vegetable dyes, as has been shown in various experiments with extract of Madder and the juices of the Pokeberry, but in no instance on record has the coloring matter produced any permanent effect on the plant or become hereditary, and it is seldom that the added color can be traced upward in the cells to any considerable distance; showing that the liquid as filtered through the cell-walls, as it passes from one to another, soon parts with any uncongenial foreign materials which may be present; at least the attempt is made to do this, and in case of failure, as with poisonous gases, the plant is killed. While the living cells may reject coloring matter and fail to retain it, the dead tissues of plants are readily colored by absorption—a purely mechanical operation—as constantly practised by the manufacturer of microscopic slides, the stainers of wood, and dyers of vegetable fabrics in general.

It is quite evident that whatever is absorbed by the roots is subjected to unceasing changes and transformations, the result of the action of chemical and vital forces about which there is yet much to be learned.

Roots that grow in the dark possess somewhat different functions from those growing in the light; at least the chemical changes which take place in them are different. We know that roots differ very widely, not only in their forms and structure, but in their habits as well. Some appear to require resistance, like those of trees which thrive best in heavy, compact clay or loamy soils; others flourish in loose sands and peat-bogs; while still other kinds grow only in water, or wholly exposed to the air, as seen in some of the epiphytes.

The food of plants consists principally of a few simple elements, viz., oxygen, hydrogen, carbon, and nitrogen. These are indispensable nutrients, out of which all combustible parts of the plant are formed by the chemical and vital processes of nutrition. It is true that other substances are usually found in plants, such as potassium, calcium, magnesia, iron, phosphorus, silicon, sodium, and various other elements, but just what position they hold in vegetable economy has not been fully determined. There are also elements which may be essential to some kinds (like iodine in marine plants) that are of no value to others.

Oxygen is a very important element of plants, for every nine pounds of water contains eight of oxygen, and it is always present in organic compounds. Plants take up oxygen, chiefly in its combination with hydrogen, in the form of water, and we all know how important moisture is to vegetation in general, as it is the vehicle which conveys to plants the great bulk of their food. Oxygen combines with various other elements to form the solid rocks of the globe, as well as the bodies of animals.

Hydrogen is an invisible element of plants and the lightest of all known substances. It is not found free in nature, but combined with oxygen in water, and it is in this state of combination that it is taken up and utilized by plants. As water is composed of eight parts (by weight) of oxygen and one of hydrogen, the latter may be considered as always present where there is moisture, and without this compound all plants soon perish. Hydrogen is always present in all organic compounds, but it is not supposed to enter into the composition of the mineral masses of the globe, but it is present in the air in combination with nitrogen.

Carbon is a constituent of every organic compound, and even the lowest order of plants, that consist only of a single cell, is supposed to have the power of decomposing and utilizing carbonic acid. On an average, forty to fifty per cent. of the weight of plants, when perfectly dry, is carbon, and in some trees and shrubs the percentage is still greater, as shown when burned for charcoal. From whence all this carbon is derived is as far from being a settled question among vegetable physiologists as is that of how it finds an entrance to the cells of plants, referred to on a preceding page. Some authors assert that it is all derived from the atmosphere through the leaves, while others are just as positive that it is taken up by the roots and then decomposed, combined, or reorganized in the cells. Prof. C. H. Goessmann, in "Manual of Agriculture," 1885, says: "that both carbonic acid and ammonia are always found in the atmosphere, and are taken in by the leaves or dissolved by the rain falling through the air and carried into the earth, where they are absorbed by the roots." Prof. Moll, an eminent German authority, so late as 1878, claims that "roots take no part in supplying the plant with carbon dioxide," while Prof. Julius Sachs, in his voluminous work "Text-book of Botany," says "that it is only the cells which contain chlorophyll—and these under the influence of sunlight—that have the power of decomposing the carbon dioxide taken up by them, and at the same time setting free an equal volume of oxygen in order to produce organic compounds out of the elements of carbon dioxide and water, or, in other words, to assimilate." The theories advanced by Prof. Moll and Sachs are also held by many other equally eminent botanists of the present day, while many of our most learned and celebrated chemists, like Prof. Goessmann, dissent, offering equally as good reasons, with some well-supported facts, for rejecting them. It is certainly consistent with the almost universal observation of practical as well as scientific cultivators of plants, that soil containing organic or vegetable matter is far more productive than that from which it is wholly absent. It is true that some kinds of plants will grow in a soil containing no perceptible amount of vegetable matter, yet we all know how much more luxuriantly plants will grow in the presence of an abundance of decomposed or decomposing carbonaceous materials. The fact, however, should not be overlooked, that in the decomposition of vegetable matter, other elements are set free in addition to carbonic acid, and it is not readily determined as to which one among the number contributes most to the increased fertility of the soil.

Every intelligent cultivator of plants knows that as each crop is taken from the land, its fertility is lessened, owing, in part at least, to the loss of organic matter; but if each successive crop derived its entire carbon from the air, and through the leaves of the plants, then it would never be necessary to add anything to the soil likely to yield carbon; consequently the theory of plants deriving all their carbon from the dioxide absorbed by the leaves is scarcely reconcilable with what appears to be the ordinary operations usually practised, if not positively necessary, in the cultivation of plants. We can readily understand, or at least believe, that it is possible, as claimed, that all the carbonaceous matter now present on this earth had its origin in the atmosphere; but it does not necessarily follow that each individual plant, or crop, derives all, or any considerable part, of its carbon from the air.

Nitrogen is another element of plants, about the origin and way in which it is utilized by plants, neither vegetable physiologists nor chemists have been able to agree. While some contend that plants cannot assimilate atmospheric nitrogen, others have proved, to their own satisfaction at least, that they do, under certain conditions, obtain considerable quantities from this source, and Adolf Mayer, in some experiments made a few years ago on air-plants, found that nitrogen, in the form of ammonia, was absorbed in appreciable amount by both leaves and roots, but most freely by the latter.

Nitrogen forms nearly four-fifths in bulk of the atmosphere, and is also abundant in all animal tissues, which, during decay, give off nitrogen, combined with hydrogen, in the form of ammonia. The latter is readily absorbed by moist carbon (charcoal), and by carbonaceous matter generally. In this form, plants take up nitrogen quite freely through their roots; consequently, ammonia is valued highly as one of the most powerful and stimulating of fertilizers. Nitrogen and the oxygen of the air, under certain conditions combine, forming nitric acid, and this, in combination with alkalies, forms nitrate of soda, of potash, and of lime, all of which are useful fertilizers for plants. The guano deposits on the islands of the Pacific Ocean, the nitre beds of South America and other countries, are all drawn upon by civilized nations for providing nitrogen and other important elements required by cultivated plants. Not only is nitrogen supplied to plants by the application of nitrates to the soil, but in various other forms of animal and vegetable manures, as produced on the farm and in the garden, and, in addition, it is being constantly deposited in the soil wherever animal or vegetable matter is undergoing decomposition.

These four elements—oxygen, hydrogen, carbon, and nitrogen—are generally recognized as the four elementary constituents of plants, supplied principally in the form of carbonic acid, water, and ammonia. In such forms or combinations they all exist in the air as well as in the earth, hence the means of subsistence of plants that live suspended in the air, as well as those the roots of which are buried in the soil.

But there are many other important elements and combinations, about which so little is known in regard to their origin or action in the building of vegetable structures that the most I need say about them is, that they are important materials and should be supplied whenever and wherever required. Among these are *Sulphur*, which is found most abundant in plants yielding what are termed albuminoids. It is especially abundant in plants of the Mustard Family, from the seeds of some plants of which is expressed a valuable oil. It is also abundant in Peas, Beans, Clover, and other seeds of legumes. Sulphur is found in the form of sulphuric acid combined as calcium sulphate or sulphate of lime, also known as gypsum and plaster. Plants take up sulphur in the form of soluble salts of sulphuric acid, but exactly how these are utilized by them is not definitely known, and it is perhaps for this reason that the use of gypsum as a fertilizer for plants has so long remained a bone of contention among agriculturists. Sometimes the results obtained from an application of this material are seen in a marked improvement in the growth of the plants, but in other instances it has no apparent effect, and this, too, on the same kinds of plants, and, so far as can be determined, on the same kind of soil.

Iron is an indispensable element of all plants containing chlorophyll—*i. e.*, with green colored parts or organs. It is, however, required in such small quantities that it is readily obtained from the soil in all parts of the world. Too much iron in the soil is injurious to plants, especially when in solutions that are readily absorbed and distributed through the cells.

Lime is an essential constituent of the ashes of plants, and it is taken up as a sulphate of lime in such plants as the Clovers, while in others, like Wheat, Rye, Oats, and similar cereals, as phosphate of lime (a compound of phosphoric acid and lime), and it is from such seed or grain that the phosphorus found in the bones of animals is produced, and without which this could not be formed. Sulphates and phosphates are necessary to supply a part of the material forming the protein compounds found in grain.

Silica is a component part of a large number of plants, and it is a combination of oxygen with a metal-like element called Silicon. Common flint and the quartz rocks are composed mainly of silica, and the transparent crystals of quartz, used for making what are called "pebble" glasses in spectacles and similar purposes, are merely a purer form of the same material. Silica, or silicic acid, is absorbed by the roots of plants largely dissolved in water as silicates, and from this solution it is deposited in the plant-cells, and in widely variable quantities. It is found in great abundance in trees, shrubs, and other woody plants, also in the bark or epidermis of the larger grasses—Wheat, Rye, Oats, Sorghum, Indian Corn, Bamboo, and the Tubular Palms; in fact, it may be termed the great stiffening material of plants. In the Bamboo it is deposited in such large quantities in the cavities of the stems that it is extensively extracted and used under the name of *tabasheer* by the Hindoos, among whom it is in high repute as a tonic. As found in the Bamboo and some of the other large grasses, it consists chiefly of silica and potash, in the proportion of about seventy parts of silica and thirty of potash. In some kinds of plants, like the common Scouring Rush (*Equisetum*), the epidermis is almost pure silex, and the ashes of the entire plant are nearly or quite one-half composed of it. Grain raised on land deficient in silica will be weak in the stem and easily blown down when the grain forms in the head or ear. As silica is found in the ashes of plants, we can readily understand how that released by decomposing vegetable matter will yield to growing crops this material in a readily soluble form.

Soda and Potash are found abundantly in all of the plants belonging to the higher orders. Those growing near the seashore usually contain a larger proportion of soda than those growing inland, while the latter contain more potash. Common potash (carbonate of potassa) is a compound of carbonic acid and potassium, while soda is a carbonate of sodium; the base of both being metals having a strong affinity for oxygen. All the alkalies—soda, potash and ammonia, and especially in their combinations with acids—form neutral compounds from which plants obtain a large portion of their mineral parts.

Oxide of Magnesium, better known under the name of Magnesia, also chlorine, iodine, bromine, alumina, manganese, and even copper in minute quantities, exist in plants, but the more important of these are found naturally in all fertile soils, and they are seldom lacking in the infertile or barren ones.

CHAPTER V.

STEMS AND THEIR APPENDAGES.

There are a vast number of simple plants that have no true stems, but are composed of only single or a multiplication of cells, and the growth of which consists merely in a division or expansion of cellular tissues. These plants do not possess a true vascular system, although, in many instances, they assume an elongated form, the cell uniting or expanding into a single filament or several parallel rows, while in others they branch out in various directions, or expand into membranes, as seen in the common lichens growing on rocks and on the old bark of trees.

In the most familiar acceptance of the term as applied to plants, a stem is that part bearing leaves and flowers. The ordinary herbs or herbaceous plants do not produce perennial woody stems, but annual flower stalks (*caulis*), which may or may not bear true leaves in addition to those organs generally accompanying the flowers.

The stems of grasses are mostly hollow, jointed tubes, living only a sufficient time to perfect their seed, whether the period required is but a few weeks, as with the common meadow grasses, or several years, as with the great Bamboo of Oriental countries. It may be mentioned, however, that while under purely natural conditions the stems of such plants invariably perish with the ripening of their seeds, it is not at all difficult to prevent either, if the stems are required for other purposes, which is often the case, as, for instance, the stalks of the tropical Sugar Cane (*Saccharum officinarum*). This plant has been so long and continuously propagated by cuttings or "rattoons," that it no longer produces seed under the artificial conditions to which it is subjected. The Bamboos, Reeds, and many other kinds of grasses, may be readily propagated in the same way, and by preventing the production of seeds they can be perpetuated and multiplied almost without limit, even when no seeds are produced. In the true Palms (*Palmæ*), the stems are perennial, and often attain to a very large size, living to a great age and fruiting almost constantly.

Among the Ferns (*Filices*), we find some with distinct fruiting stalks, the seed-like organs being produced on leafless, sporiferous stems, springing direct from the crowns of the roots, while in others the sporangia are found on the fronds only. Plants producing distinct stems are termed caulescent, while those in which the stem is inconspicuous are acaulescent—*i. e.*, without a stem. Botanists separate flowering plants into two great divisions—the *Endogens* and the *Exogens*, or "inside growers," and "outside growers." As these terms have a special reference to the growth and structure of the stems of plants, it is proper that they should be explained here; although the differences between the plants of the two divisions are usually distinguishable in the seed as well. Endogenous stems are not made up of concentric rings or annual layers of deposited matter, as seen in the woody stems of nearly all exogens or outside growers. In the formation of the woody tissues of endogens the new material deposited appears to be intermingled with the old, and the increase in the size of the stems is principally through distention or pressing outward, and not by the deposition of matter in the form of layers, such stems consisting of bundles of fibers intermingled with or imbedded in cellular tissues. Neither do such stems show the marked distinction between the pith, wood and bark, as seen in those of exogens. The Palms, Ferns, Yuccas, Bamboo, Sorghum and all of our cereal grasses belong to this division of true inside growers. Their seeds are also distinguished by having only one cotyledon or seed-leaf, hence are called *monocotyledonous* plants, the plumule pushing upward from the seed in a columnar form, as seen in the Asparagus, Indian Corn, or the giant Palms of the tropics. In the leaves of these plants we also find that the veins run mostly parallel with the length—that is, extend from base to point and not branched.

The exogenous stem has at first three distinct parts, viz., the pith, wood and bark, all readily separable. As the stem increases in size through the deposition of new matter in concentric layers of bundles of wood-cells, the pith is often compressed or entirely obliterated without in any manner interfering with the growth of the plant, for the principal office of the pith is to facilitate the rapid transmission of fluids through the succulent stems of herbaceous plants and the young plantlets, and twigs of shrubs and trees. The solidity of the stems of the trees diminishes from the center to circumference as they increase in size, or just the opposite of what takes place in endogenous stems, the inner portions, in time, ceasing to take any active part in the movement of the outer layers, and the heart-wood may decay, as seen in thousands of instances in almost every old forest, or be forcibly removed without severely checking the growth of the younger parts of the tree. But so long as the center of the stem remains entire, there is a slight communication between the outer and adjacent parts through what are called the medullary rays, which are composed of cells spread out into a thin membranous structure. These rays are quite abundant and conspicuous in the wood of the Beech, Maple, Oak and many other kinds of trees. Through these rays the inner parts of the stem are supposed to be supplied with sufficient moisture to prevent complete exsiccation of the wood.

Following in the same direction, from the pith outward through the mature wood, we next reach a later formation composed of a few or many concentric layers which, as a whole, are called alburnous wood, or *alburnum*, from the Latin *albus*, white, because in most trees and shrubs this recently formed wood is of a whitish color, or at least lighter in color than the *duramen* or heart wood, and the cells of which this alburnum is composed are capable of transmitting living, organizable matter. The cells of the inner layers of alburnum are, however, less active than those of the outer, and the propagator of plants by division, finds the latest formed wood responds most readily to his wishes in the various operations to which it is subjected. Encircling the alburnum we find a layer of soft organizable matter which has received the name of *cambium*, which is quite abundant in some

STEMS AND THEIR APPENDAGES. 55

plants just before and at the time they commence their growth in spring. It is of a mucilaginous nature, and filled with cells that are actively assuming their more regular and solidified form, as found in the completed or mature wood.

Outside and resting upon the cambium layer, or partly immersed, as it were, in it, we find the *liber,* or inner bark, which in some kinds of trees, like the Beech, is of a granular structure and very brittle, while in others, like the Papaw, Persimmon, and Lindens; it is cloth-like and filled with strong, tough fibres. Surrounding the liber we find the older layers of bark partaking somewhat of the character of the liber, but generally quite porous, coarser, and of a more brittle texture. The old bark of trees often breaks up into deep furrows, as on the stems of the Chestnut and Elm, or cleaving off in thin irregular plates, as from the Plane or Buttonwood tree, while in some it peels off in in the form of annular paper-like rings, as in the Birch and Cherry. This outer bark is merely effete or dead matter, an excrescence of no further use to the tree than to cover and protect the inner bark from the elements.

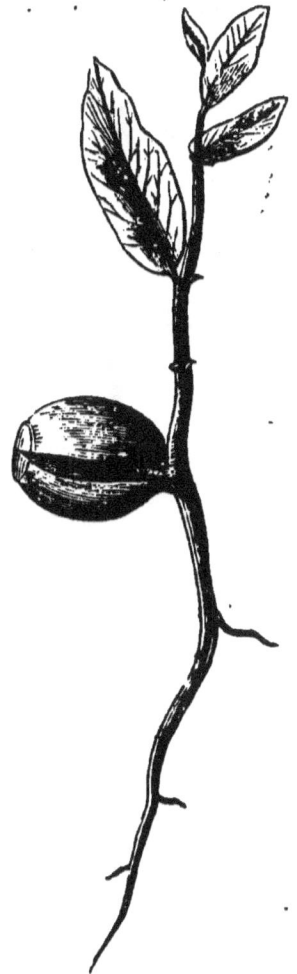

Fig. 14.—SEEDLING OAK.

The exogens are also called dicotyledonous plants, because their seeds have two *cotyledons,* or seed-leaves, as has been explained elsewhere, but may be referred to again for the purpose of noting the exceptions to the rule, as well as to explain more fully some peculiarities of stems in their embryonic stages.

In the young seedling Oak, shown in figure 14, the two cotyledons or seed-leaves are attached to one side of the stem and remain partly enclosed within the shell, this being the usual position in which they are found on the starting plantlets, for they seldom expand sufficiently to free themselves from their horn-like covering; neither is it necessary, for the secondary leaves are early developed, and before the nutriment stored up in the nut is entirely exhausted in producing the plumule and simple root. A similar restricted development of seed-leaves occurs in nearly all of the nut-like seeds, and even in the seeds of many herbs.

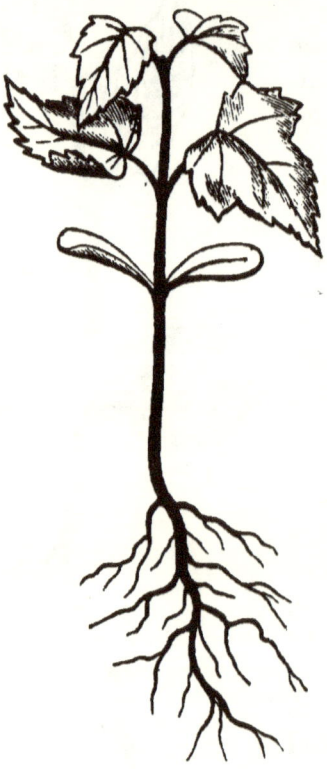

Fig. 15.—SEEDLING MAPLE.

The common garden Pea is a familiar example of a dicotyledonous seed, the seed-leaves of which do not "come up," or appear above-ground on the ascending stem; but on the closely allied garden Bean the cotyledons or seed-leaves are always conspicuous objects on the young plants. This is also true with many kinds of trees, such as the Pear, Apple, Plum and Cherry among fruits, and the Ash, Elm and Maple among our most familiar forest trees; as soon as the second and third pair of the true leaves unfold, the seed-leaves below commence to dry up, as shown in the seedling Maple, figure 15, and eventually drop off.

The position as well as the movements of cotyledons in seedlings are characters worthy of careful study by cultivators of plants, for it is quite important to know in advance of sowing seeds whether the first leaves are to "come up," as with the Bean, or remain stationary below the surface, as with the Pea, Wistaria, Oak and similar seeds, for in the former the cotyledons must break through the soil and come to the surface, while in the latter it is only a new and slender stem therefrom that appears. Different conditions, therefore, are required for different growths, for the large, fleshy seed-leaves of some kinds of plants would never break through a compact soil or crust on its surface that might not seriously impede the progress of plants with cotyledons, which remain stationary at the depth at which they are deposited.

While the two cotyledons are generally recognized as a characteristic of plants with woody, exogenous stems, still there are some exceptions; among the most familiar of these are the conifers, or cone-bearing trees, for in these the cotyledons or seed-leaves are quite variable in number. In the seedling Arbor-vitæ (*Thuya*), the usual number is two, but in the Pines they range from four and five up to fifteen or sixteen in Sabine's Pine (*Pinus Sabiniana*.) The seed-leaves in the Pines are produced in a whorl, as shown in figure 16, and they always push their way up above the surface, if not prevented by a too compact soil. The true leaves of the Pine tree grow in clusters, or, more properly, bundles, the lower ends being encased in a kind of sheath.

The number of leaves in a sheath varies in different species from one or rarely two, in the One-leaved Pine (*Pinus monophylla*), up to five in the common White

Pine (*P. Strobus*), figure 17. The cotyledons, however, give no indications of the number or arrangement of the true leaves, which appear later on the plant, for in the One-leaved Pine there are from seven to ten cotyledons, while the seeds of Sabine's Pine produce fifteen to sixteen, and later the true leaves are arranged with only three in a sheath or bundle.

Another dicotyledonous character is wanting in the leaves of a large majority of the conifers, and that is the branching veins, for, with only a few exceptions, their leaves are long, slender and with parallel veins; consequently, in seeking characters to aid us in separating the dicotyledonous from the monocotyledonous plants, we must not expect to find all equally well developed, or even always foremost in any one genus or family. For instance, the common garden Pea has two distinct and readily separated cotyledons and its leaves have branching or netted veins, but the stalk shows no outside growth, this character being confined wholly to stems that live more than one season, and it is never developed until the second year. The Wistaria vine is closely related to the Pea, but the stem being perennial, new layers of wood are annually added to the outside; consequently, it has a true exogenous stem.

Fig. 16.
SEEDLING PINE.

Fig. 17.—LEAVES OF WHITE PINE.

The appendages of stems are exceedingly numerous and of diversified form and structure, and all are of value, and of more or less importance in aiding us to distinguish plants of the different classes, orders, genera, species, and even the natural and cultivated varieties; but the space at my command will only admit of a brief notice of the few with which the propagator must necessarily become the most familiar. I will say, however, that he who aims *to know* plants must not think that even the minutest character is unimportant, for size is only a comparative term at best, and a thing may be great among the small as well as among the large. The most prominent appendages of stems are prickles, as in the Nettle; spines, as found on the canes of the Black Raspberry, the Blackberry and the Roses; thorns, as on the Hawthorn, the Honey Locust, and many of the larger cactuses; tendrils, as on the Clematis, Grape, and many other climbing plants. Leaves, flowers, fruits or seeds are other appendages. Prickles, spines and thorns probably take an active part in the general assimilation of nutrients of the plant, at least while they are young and growing, but what other purpose they serve in vegetable economy is not readily determined, further than they are distinguishing characteristics among the vast numbers which Nature employs in her always different and ever changing productions. To say that plants are armed with spines or thorns as a protection, as is often asserted, has no foundation in fact, but it is a purely sentimental idea, for the supposed protecting organs do not protect against any natural enemy, for the species most fully armed with the strongest spines and thorns often perish from the attacks of some thin-skinned and wholly defenseless little insect, while the giant thorns of some trees often become the safe and rather luxurious home of certain species of the ant.

There are other appendages of stems which may not

take any very active part in gathering or assimilating nutrients, and still be of great service in other ways to the plants producing them—for instance, the filiform aerial organs on the stems of the Poison Sumac (*Rhus Toxicodendron*), on the Trumpet Creeper (*Tecoma radicans*), American and Japan Ivies (*Ampelopsis*), and many other similar and well-known plants. These ap-

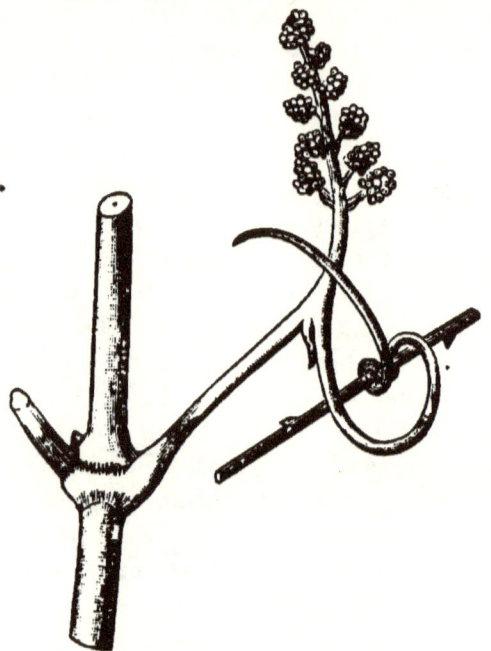

Fig. 18.—GRAPEVINE TENDRIL.

pendages differ much in their mode of attachment to whatever object serves them as a support. Some force themselves into the cracks of the bark, wood, or rocks, or are attached by minute lateral fibres, while others, like the Ampelopsis, produce small flattened discs on the ends of their many-branched, root-like organs each disc becoming fastened to whatever object that happens to be within reach. Tendrils are merely modified forms of the same organs in these and allied plants, possessing various morphological characters, often changing and serving different purposes. In many species of plants having woody stems, like the Grape, Passion-flower and American Ivy, the tendrils are really metamorphosed flower-stalks, for while the larger proportion serve in assisting the plant to climb and retain a position where the leaves will be exposed to the light, a much smaller number on the same plant may blossom and eventually become a bunch of fruit; that is, a bunch of fruit on such plants is merely a productive tendril. It is only a few—from one to five—of the tendrils first formed on the young Grape-canes of the season that are fruitful, all that are produced later being unproductive or barren, possessing great irritability, which causes them to cling to or twine about any object with which they come in contact. The ends of the tendrils of the plants under consideration are divided, sometimes into several branches, as in the American Ivy, or into two or more, as in those of the Grape, two being the most usual number in the latter, and these not of the same length; consequently, when both divisions are fruitful, the bunch of grapes will be double, with one side shorter than the other, the lesser bunch having the technical name of "shoulder." If there are two short and fruitful branches the bunch may be double-shouldered, or if there is a greater number the bunch may be a cluster. Such terms as single-shouldered, double-shouldered and clustered bunches are employed by pomologists in describing the form of the bunch in the different cultivated varieties of the Grape.

It very frequently occurs that only one division of the tendril will be fruitful, as shown in figure 18, the other remaining barren, winding around some convenient branch or twig. That the tendrils are of the same nature as other parts of the plant, the juices flowing through them as actively as in the stem, is shown by the fact that a fertile may be grafted upon an unfertile tendril. As it is a characteristic of the tendril to turn away from the light and seek the shade, it naturally follows that the fruit of these plants also ripen best when protected from the direct rays of the sun, as is well known to every practical cultivator of the Grape.

In many herbaceous plants the tendril is but a prolongation of the mid-rib beyond the point of the leaf, as seen in the Pea-vine, and, in a few instances, like that of the Yellow Vetchling (*Lathyrus Aphaca*), of Great Britain, the whole leaf is but a filiform tendril, while in such climbers as the *Clematis*, *Maurandia* and *Lophospermum* the petiole of the leaf may serve as a tendril.

All twining plants may be considered in the nature of tendrils, being irritable and sensitive on one side, enabling them to climb supports and retain an upright position, but the biology of such plants is scarcely of sufficient importance to the practical horticulturist to call for treatment in detail in a work of this kind.

Fig. 19.
LEAVES OF JERSEY PINE.

Buds may be placed in the list of appendages of stems, for they are extensively employed in the propagation of plants, being removed and transferred from one to another with a portion of the surrounding bark and wood attached, and, in such positions, becoming a part of the stem to which they are united. They are also, in some instances, placed in a position where they produce roots, and thus become separate individual plants. Buds may therefore be briefly described as organs enclosing within scales the rudiments of a stem, of leaves or of flowers. It naturally follows that the appendages of highly-developed plants, which are called leaves, are merely the unfolding of buds and a combination of the tissues of the stem or other parts from which they are developed. Leaves are generally formed by the elongation and expansion of the ligneous bundles of tissues, the interspaces being filled with cellular matter (*Parenchyma*), of green or greenish color.

That the woody part or frame-work of the leaf is of

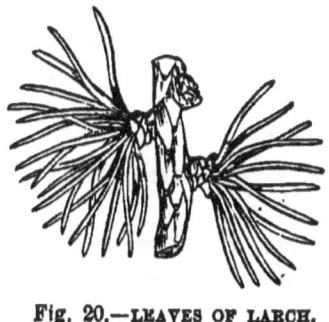

Fig. 20.—LEAVES OF LARCH.

Fig. 21.—LEAF OF LILAC.

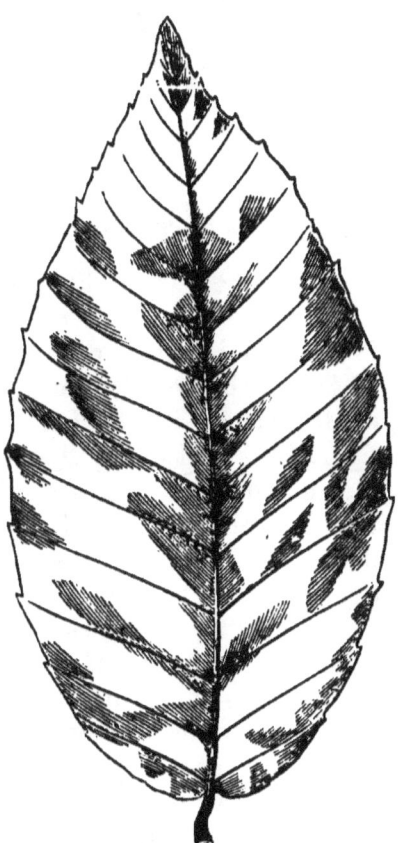

Fig. 22.—LEAF OF BEECH.

the same nature as that of the branch or stem is shown in the readiness with which many kinds produce both buds and roots; consequently, leaves may produce stems as well as stems leaves.

The general form of the leaf depends mainly upon the disposition of the principal veins and branches of the woody tissues of which the skeleton of the leaf is composed. When these tissues run parallel, and are composed of a single or several thread-like fibres, the leaves will assume a similar form, as seen in the linear-shaped leaves of the Pine (figure 19, Jersey Pine, *P. inops*). The veins in these leaves starting singly from the stems, are each surrounded or incased in cellular matter. This single form of growth is common in many of the conifers; the leaves, instead of having branching veins, grow in a thread-like bundle or fascicle, as in the Larch, figure 20. But in the simple membranous leaf, like that of the common garden Lilac, figure 21, the central stem of the leaf, for about one-third of its length, forms what is called a *petiole* or leaf-stalk; then throws out branches, all remaining united by the thin membrane or parenchyma, which fill the interspaces. In some leaves, like those of the Beech, figure 22, the secondary veins branch off at an ascending angle from the mid-rib, running almost in a straight line to the outer edge of the leaf, forming a saw-tooth-like notch where each terminates. The edges of such leaves are said to be *serrate*, because beset with teeth projecting forward like those of a saw. But in

Fig. 23.—LEAF OF CUT-LEAVED BIRCH.

Fig. 24.—LEAF OF BUCKEYE.

STEMS AND THEIR APPENDAGES. 65

the leaves of the Cut-leaved Birch, figure 23, the branching veins are of irregular or unequal lengths, giving to

Fig. 25.—LEAF OF LOCUST.

Fig. 26.—LEAF OF ACACIA GRANDIFLORA.

the edges of the leaves a jagged or tooth-like form. In the leaves of the Buckeye, figure 24, the mid-rib branches into five parts, each division having the ap-

Fig. 27.—LEAF OF FERN-LEAVED ARALIA.

pearance of a simple leaf, but all attached to the petiole at one point. Leaves of this kind are called *palmate* or *digitate*, because they resemble or are divided like the fingers on the hand.

In the leaves of such trees as the Butternut, Black Walnut and common Locust (*Robinia*, figure 25), the divisions of the leaf are scattered in pairs along both sides of a greatly lengthened mid-rib, and such leaves are said to be *compound* or *pinnate*, from the Latin *penna*, a feather. A further subdivision of the leaflets in this kind of a leaf is seen in the *Acacias*, and a leaf of *A. grandiflora*, much reduced in size, is shown in figure 26. Leaves divided in this form are said to be *bipinnate*, or twice pinnated. There are many hundreds of different forms of these bipinnate leaves, as well as of all other forms, and in the *Aralias* and closely allied plants the leaves are often of immense size. In the Fern-leaved Aralia (*A. filicifolia*), the upper part of the leaf-stalk expands into a broad, leafy branch, which is curiously divided, as shown in figure 27.

CHAPTER VI.

FLOWERS, FRUITS AND SEEDS.

Flowers, like all other organs and appendages of the stems of plants, assume almost innumerable forms and sizes. Some are almost microscopic, or remain concealed within an envelope, as in the common Fig, while others are large and very showy, as in several species of the Magnolia; consequently, it is difficult to give any concise and clear definition of a flower that will apply to all. In the ordinary acceptance of the term, a perfect flower is one that contains all the organs of reproduction, as seen in that of the Apple, Rose, and many other similar and common plants. At the base and outside we find a floral envelope or *calyx*, and this is divided into several leaf-like divisions called *sepals*. Within the calyx there is another set of floral organs called *petals*, which in the flowers named, are larger and broader than the sepals of the calyx, and usually quite conspicuous—often of a brilliant color, and together form the *corolla*. The next row of organs are *stamens*, these being slender and thread-like, composed of a stem or *filament* tipped with a knob-like body, called an *anther*, the latter being filled with a fine powder, *pollen*, that differs widely in size and form in different kinds of plants. The stamens are really the male organs of plants, for the pollen produced by them is the fructifying substance, without which no seed can be formed in any of the higher orders of plants.

In the very center of the flower we find another set of slender organs called *pistils*, which rest upon or are but a part of what is called the *ovary*—the organ that contains the *ovules* or embryo seeds. The pistils are the female organs, and it is through these that the pollen from the anthers exerts its influence upon the embryo seeds in the *ovary*. The upper or extreme point of the pistil is called the *stigma*, and the column, or stem below, the *style*, that widens out at the base into an ovary. But these simple and perfect flowers are but one among the many thousands of forms to be found everywhere among plants, and while in nearly all of the plants with distinct flowers the pistils and stamens can be readily distinguished, their size and position are far from being uniform. In simple and perfect flowers, like those of the Apple, both sets of organs are found in the same flower; consequently, they are called bi-sexual or perfect, but in many other kinds of plants, these organs are found in separate flowers on the same plant, as seen in the Pines,

Oaks, Beech, Chestnut, Walnut, and Hazel. Such plants are called *monœcious*, because each produces but one kind of sexual organs. In other plants the staminate and pistillate flowers are produced by separate plants, as in the Poplars, Buffalo-Berry, Hop and Hemp, and in some of the Maples. The flowers of such plants are said to be *diœcious*, because the two different sexual organs are borne by separate individual plants. There are also species of plants distributed among various families and genera, like certain species of the Grape, Ash, Maple, Olive, and many of the Palms, which bear flowers, some with pistils only, others with stamens, and some with both kinds of organs in the same flower. Plants with these variable flowers are said to be *polygamous*.

In plants like the Asters, Gaillardias, Heleniums, and the common Sunflower, the flowers are called compound, being crowded together in a broad head; the position and distribution of the sexual organs are variable (*heterogamous*), some containing both stamens and pistils, while others have neither, and are therefore neutral or abortive, as often seen in the ray florets or outside rows of showy petals in such flowers.

But it is not my purpose to attempt to describe or even note the many forms and the variability in the structure of flowers, but merely to call the attention of the reader to the fact that such variations not only exist, but require close and careful investigation by persons who desire or intend to become successful cultivators and propagators of plants. It must be apparent to the most casual observer that it would be perfectly useless to set out one specimen of our native Buffalo Berry (*Shepherdia*), expecting it to produce fruit, for the sexual organs are in separate plants, and the one bearing staminate flowers never produces fruit under any circumstances, and the plants bearing pistillate flowers will not bear, except in the presence of the staminate; consequently, it is imperative that at least one plant of each sex shall be growing near enough together to admit of the pollen from the anthers to reach the stigmas of the pistils with the aid of the wind or insects. One staminate plant may be sufficient to fertilize the flowers of several pistillates, but the two sexes must always be present in order to secure fruit and seed. The same rule holds good in all dioecious trees, shrubs and herbaceous plants. For further information on this subject, I would refer the reader to the various standard works on "Structural Botany."

Fruit, in the ordinary meaning of the word, is something different from seed, although, in a strict botanical sense, they are really inseparable, for in many plants it requires both seed-vessels and seeds to make a perfect fruit, while in others the fruit includes other parts of the flower, such as the bracts and floral envelopes. But the most common idea of a fruit is something edible and different from the seed, although it must be said that the terms edible or eatable are rather vague and indefinite, inasmuch as a fruit may be eatable for one kind of animal and not another, and it may also be edible, palatable and healthful for mankind in its improved condition, while the same species may be neither of these in its wild or primitive condition.

In the Peach it is the juicy, agreeable pulp surrounding the stone, or seed, that becomes the edible part, but in the Almond, which is closely allied, if not the actual parent of the Peach, the outer envelope is dry and not edible, the kernel or seed alone being considered valuable as food.

In common horticultural phraseology, fruits are separated from nuts, grain and other kinds of seeds, not that the division is always scientifically correct, but such classification is made as a matter of convenience, when speaking of the members of each class. For instance, when we speak of "small fruits" or berries, it is not to be supposed that they are all really small, or, scientifically speaking, true berries, but by common consent the trailing plants, like the Strawberry, Cranberry, Blackberry, or upright growing bushes, like the Raspberry, Gooseberry and Currant, have received the name of small fruits or berries.

There is such an immense number of fruits, and they are so variable in structure that I can only refer to a few, merely to show, in a very general way, their manner of growth. There are simple and compound fruits and various sub-divisions of each class. The common Plum (figure 28), Peach and Cherry are familiar examples of what are termed simple fruits, or the ripening of a one-celled pistil, the seed or kernel being surrounded by a hard, bony shell, and this enclosed in a fleshy, edible pulp. In the Cornel (*Cornus Mas*), and common Dog-wood, the seed is composed of two bony cells, one often abortive, but all surrounded by the fleshy, edible pulp. The olive is also a *drupe* or stone fruit, but with a one-celled seed vessel.

Fig. 28.—COMMON PLUM.

The Raspberry and Blackberry are really an aggregation of small stone fruits, their hard, bony seed being surrounded with an edible pulp, each seed being a ripened pistil, but all arranged on a conical or elongated receptacle. But in these fruits the entire cluster is the product of one flower. In the Mulberry, however, which so closely resembles the Blackberry in form and general appearance, the fruit is really a *Sorosis* or congeries of fruits, the product of numerous female or pistillate flowers united; the calyx of each becoming succulent and adhering to the ovary. The Bread fruit (*Artocarpus*) and the Pineapple (*Ananassa*) are multiple fruits of this kind, formed by numerous ovaries, floral envelopes and bracts combined, all uniting and becoming a succulent mass.

The common Fig, although a multiple fruit, is quite the opposite of the Mulberry, and is a *Syconus*, the peduncle or fruit stalk, becoming hollow, bearing the numerous minute flowers within the cavity, where all are united, producing the flesh, or what is usually termed the fruit.

Fig. 29.—STRAWBERRY FLOWER.

The Strawberry has the appearance of a Fig turned inside out, but, instead of being the product of many flowers, it is of only one, with many pistils, as shown in figure 29. The petals and stamens drop off, leaving a central fleshy receptacle resting upon, or attached to, the apex of the peduncle. As this fleshy receptacle enlarges, the ovaries or seeds spread apart, either becoming slightly imbedded in or resting on the surface of the mature part, as shown in figure 30.

In the Pomeæ — Apple, Pear, Medlar, Mountain Ash, Hawthorn and Quince—there are from two to five cells, with thin or thick walls, enclosing one or two seeds in each cell in the Pear and Apple, or several as in the

Fig. 30.—STRAWBERRY.

Quince. The seeds are mainly enclosed in a thin membranous covering, but in the Hawthorn and Medlar it is hard and bony. In the formation of the true pome the calyx tubes enlarge or thicken, becoming a soft and pulpy fruit; the sepals or leaf-like divisions of the calyx are carried forward as growth proceeds, and remain attached, as seen in what is called the blossom end of such fruits. In the Gooseberry and Currant a somewhat similar enlargement of the calyx takes place, but the ovary is only one-celled, the seeds being imbedded in the pulp, and attached to the two opposite sides of the cell.

The Orange and Lemon are berry-like in structure, the ovary free and many-celled, each containing one or more seeds, with thick, fleshy cotyledons. These fruits have thick, spongy rinds, and pulpy separable cells, the cell-walls thin and membranous. The Persimmon (*Diospyros*) is also a fruit of a berry-like structure, with large bony seeds imbedded in pulp. The cells of the ovary are irregular, and the styles and stigma united as one, but with several distinct pollen tubes.

The Papaw or Custard Apple is a large plant with dull-colored flowers of six petals in two rows, succeeded by large, oblong, pulpy fruit, containing several large, flattish, long seeds. There are, however, many species belonging to this family, and the fruit is quite variable in form and structure. But, as I have already stated, the number of different kinds of fruits is far too great to admit of even a brief description of all, or any considerable number of them, in a work of this kind; for even in our cool climate there are, in addition to those already named, the Grape (*Vitis*) in many species, the Barberry (*Berberis*), Buffaloberry (*Shepherdia*), Blueberry (*Vaccinium*), Low Trailing Cranberry (*Vaccinium Oxycoccos*), High Bush Cranberry (*Viburnum Opulus*), Elderberry (*Sambucus*), Huckleberry (*Gaylus-*

FLOWERS, FRUITS AND SEEDS. 73

accia), Mandrake or May apple (*Podophyllum*), Partridgeberry (*Mitchella*) and Wintergreenberry (*Gaultheria*),
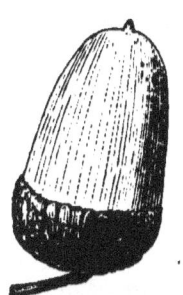
besides Melons, Squashes, Pumpkins, Cucumbers, Tomatoes, etc., all of which come under the head of edible fruits. In tropical and semi-tropical countries there is still a greater number which I must omit mentioning here, but nearly all will be named elsewhere.

NATURE OF SEEDS.—Seeds are embryo plants, the mature, fecundated ovule, with certain nutritive properties stored up within various protective organs. A seed is also a living organism which separates from its parent, and is then capable of becoming a new individual of the same species. When mature they

Fig 31.
Q. PEDUNCULATA.

contain various albuminous, ligneous and oily compounds required to supply the young plantlet with nutriment during its early stages of growth, or until roots are produced through which it can obtain nutrients from surrounding elements.

Some seeds acquire a stony hardness when ripe, as seen in the Ivory

Fig. 32.
Q. P. FASTIGIATA.

Nut; others remain soft and fleshy, as in the Horse-
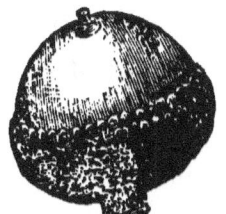
chestnut, or partly fleshy and partly liquid, as in the Cocoanut; others, like those of the maples and elms, are soft, and the cotyledons leaf-like; others are of a starchy nature, as in Oats, Buckwheat and the Onion.

Seeds are usually enclosed in a pericarp or seed-vessel, which offers protection to the kernel and germ. Sometimes there are several of these protective organs, as seen

Fig. 33.
Q. TINCTORIA.

in the Chestnut, Filbert and Walnut, the outer husk enclosing the nut becoming detached at maturity from the parent plant, as well as from the nut proper, while within the latter there is another integument in the form of a thin membrane surrounding the kernel. In some

Fig. 34.—Q. ALBA.

of the Oaks the outer husk is but a shallow cup, figures 31 and 32; in other species the cup extends farther up or is deeper, figures 33 and 34, while in a few the nut is nearly covered with the husk.

The stalk of the seed is called the *funiculus*, and when a seed breaks loose from the stalk at maturity it leaves a scar, as seen in the garden bean, which is called the *hilum*. Sometimes this funiculus is extended or even rolled up within the seed-vessel, forming a thread-like attachment to the seed, as seen in those of the Magnolias, allowing the seed to drop out and remain for a

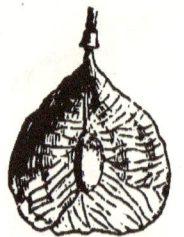  

Fig. 35.—ULMUS SUBEROSA.   Fig. 36. U. CAMPESTRIS.   Fig. 37. U. EFFUSA.

time suspended to the seed-vessel. On the seed of the Common Swamp Magnolia (*Magnolia glauca*), this umbilical cord is an inch or more in length, and quite conspicuous as the seed drops out of the cone-shaped pods in autumn.

In form, size and structure, as well as in chemical composition of seeds, variation is the rule, as well as in other parts and appendages of plants. Seeds of differ-

FLOWERS, FRUITS AND SEEDS.

ent families, genera and species may have sufficient general resemblance to enable the botanist to determine very nearly where they belong, still they will usually vary considerably in many of their characters, even when very closely related, and as near alike as two peas from the same pod, and yet no two peas are *exactly* alike.

DISTRIBUTION OF SEEDS.—Nature has provided various methods for the distribution of seeds, thereby in a measure preventing overcrowding of plants, although it is quite evident that, in her prodigality, she produces a far greater number of seeds than can possibly grow, with

Fig. 38.—SEED OF THE ASH.

room for the plants to reach maturity; but as animals are dependent upon plants for support, seeds are largely consumed, and yet, when this demand has been fully provided for, there is still a large surplus of some kinds, and the war of races, as well as of kinds, takes place, ending in what has been aptly termed "the survival of the fittest."

The seeds of many herbs and trees are provided with long hair-like appendages (*pappus*), as seen on the seed-vessels and seeds of the Thistles (*Cirsium*), Milkweeds (*Asclepias*), Willow-herbs (*Epilobium*), Dandelion (*Taraxacum*), Cottonwood (*Populus*), which assist in their distribution by the wind. Others have hooked awns on the outside of their seed-vessels, as on those of the Burdock (*Lappa*) and Beggar-ticks (*Bidens*), which become attached to any rough surface like the hide of animals or wool of sheep, and by such means become widely distributed. The seeds of many kinds of trees are provided with wing-like appendages, those of the Elm having a thin membrane passing nearly or quite around the edge of the seed, as shown in figures 35 to 37. The Ash tree has long, slender one to two celled seeds with a wing on two sides and the upper end—figure 38. The seeds of the Maples are produced in pairs, each pair the product of a single flower, the membrane of the wing growing inward from its stem, as shown in figure 39, which represents a pair of the seeds of the Large-leaved California Maple (*Acer macrophyllum*) of natural size.

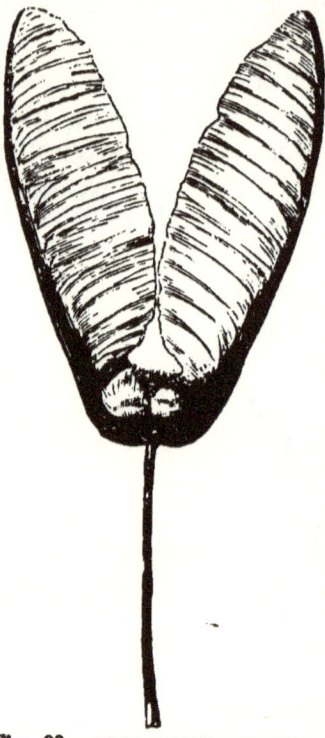

Fig. 39.—ACER MACROPHYLLUM.

In the *Coniferæ*, or cone-bearing trees, the seeds of a large majority of the species are winged, but in only a limited number are these appendages of any assistance in their distribution. In the seeds of the Great Tree of California (*Sequoia gigantea*), the Arbor Vitæs (*Thuya*), and closely allied species, the wings are firmly attached to the seeds, consequently aiding in their extended distribution as they fall from the cones, but in many species of conifers the wings are deciduous, readily parting from the seed as soon as it is mature, and before dropping from the cone. In the Pines especially, we find a wide variation in this respect, for in some species, like those of the Northern Pitch Pine (*Pinus rigida*), figure 40, the wing of the seed is three or four times the length of the seed, and it adheres until the seed has been widely scattered, but in those with larger seed, such as the Nut Pines (*P. edulis*) and others, the wings become detached before, or at the time, the seeds fall from the cone.

While wind and water, birds and mammals of various kinds, assist in the distribution of seeds, there is no discrimination in these natural processes, and the seeds of worthless and pestiferous plants are just as likely to spread over the earth's surface and grow as the most valuable and useful.

Fig. 40. SEED OF PINUS RIGIDA.

VITALITY OF SEEDS.—Perfect and mature seeds, as I have said elsewhere, are living organisms, within which there can be no absolute cessation in the action of the living principles or vital forces, for when this occurs the seed dies, and a purely chemical change follows, however slow and protracted it may prove to be. Sometimes the vital actions may appear to be almost suspended, either under normal or abnormal conditions, but they are never absolutely quiescent, and while some kinds of seeds may retain their vitality for many years under favorable conditions, there is a limit to individual life even among seeds. No known species will remain alive for "ages," as has been so often asserted by persons whose imagination seems to have been far more fertile than their knowledge was profound. It has been frequently claimed that Wheat, and other kinds of grain, taken from the Egyptian tombs and the wrappings of mummies, where it was deposited many thousands of years ago, has been made to germinate and grow when planted, but there are no well authenticated instances of any such growths. Whenever any of these supposed-to-be very ancient seeds germinate, we may conclude that there has been some trickery about the matter. Travelers in search of curiosities are usually accommodatad by servants and guides, for even Indian Corn or Maize, unknown in Europe before the discovery of America, has been furnished travelers, among other seeds claimed to have been found in the ancient Egyptian tombs. We may also distrust all accounts of the germination of seeds that have been buried in the earth for hundreds and thousands of years, and brought to the surface in digging wells, canals and other similar deep excavations, for in no instance that has been traced to any trustworthy source have the plants raised from such seeds been specifically different from those recently introduced or indigenous to the same locality. No new genera or species from such seeds have ever been brought to light, as might be expected if they were the product of some very ancient plants or locally extinct species, as usually claimed. It is certainly true that no one knows how long some kinds of seeds may remain sound when buried deeply in the earth, or for how many years some kinds may retain their vitality under the most favorable conditions, but when we come to test seeds by actual and carefully conducted experiments, we find little to support the theories advanced in regard to long duration of vitality.

Many hundreds of experiments have been, and are still being made, for the purpose of determining the limits of life in seeds, the results differing but slightly, or no more than might be expected from the unavoidable variableness of conditions, while all do show that vitality decreases more or less rapidly with age.

Size has no influence on the vitality or keeping properties of seeds, for the Cocoanut, under the most favorable conditions, will not remain alive half as long as the seeds of the common Field Poppy, which are a million of times less in bulk.

Neither has there been as yet discovered any peculiar property of seeds that would make those possessing it insensible to the influence of surrounding elements and conditions. It might readily be supposed, upon general principles, that seeds containing a large amount of oil would be far less likely to be affected by moisture, dryness, or even a low temperature, than those of an opposite nature, but long experience with such seeds shows that oil is not a preservative property. For instance, such large seeds as the Butternut, Hickory-nut, Black Walnut and European Walnut contain a large, fleshy, oily kernel (figure 41), which, from appearances, we might suppose would be able to resist ordinary adverse conditions for a long time, but, on the contrary, they are quite sensitive to extremes of any kind, and it is with difficulty such nuts can be preserved alive and sound for even a twelvemonth. If kept warm, a chemical change takes place, and the oil in the kernel becomes rancid. In a moist position the kernel soon decays, and, while a freezing temperature will check decay, if it be long continued the germ of the seed is destroyed. But, on the contrary, the minute oily seeds of the Mustard will withstand considerable drying, a high temperature or a low one, and still retain their vitality for several years. Such minute seeds as the common garden Pursley or Purslane (*Portulaca*), will withstand great extremes of temperature, also alternate soaking and drying while buried in the soil, and yet survive these changes, while the great hard-shelled nuts, as well as many of the seeds of our forest trees, must grow, if at all, or die within a few weeks, or months at farthest, after reaching maturity. There are,

Fig. 41.
KERNEL OF WALNUT.

of course, certain families of plants, the seeds of which possess much greater tenacity to life than others, but in all, time is a weakening element. The seeds of Melons and Cucumbers may be preserved in good condition for growth from one to fifteen years, and even at the latter age yield a fair proportion of plants, while the seeds of the Onion, Carrot and Celery soon perish if kept under the same conditions, few germinating after they are three or four years old.

Then there are other kinds of seeds which require age and long exposure to heat and moisture to soften their horn-like covering, as in those of the *Cratægus, Mespilus, Cornus, Halesia, Ilex,* and several species of the Rose; but in all of these, and others of like character, a few weeks' exposure to a dry atmosphere will either destroy vitality, or so harden the shell that the pores of the latter will not open again to admit moisture to the kernel.

It might readily be supposed that the natural surroundings of vessels enclosing seeds would not only afford the best protection, but also insure the most favorable conditions for the prolongation and preservation of their vitality, and while this is true in many instances, it is quite the contrary in others. The pulp of the Apple, Pear, Quince and similar fruits, if left to decay about the seed, is almost certain to destroy it, from the chemical changes which occur in the decomposition of what is called the fruit, consquently to remove the seed by artificial means is assisting nature in its preservation.

There are seeds which naturally begin to grow while still attached to the fruit-stalk, not separating from the parent plant until after germination and the production of roots. The fruit of the Mangrove tree (*Rhizophora Mangle*), of Southern Florida, is a well-known example of this kind of seed. Then, again, there are other seeds which will remain sound for many years if kept sealed up in their natural and multiple integuments, as seen in many of the conifers, and especially among the Pines and Cedars, the cones of some of these not opening or allowing the seed to escape for several years after maturity. The cone of the Cedar of Lebanon (*Cedrus Libani*), is a somewhat remarkable example of this kind, and in one instance it is reported upon trustworthy authority that seeds from a cone that had been kept in a cabinet forty years, germinated quite freely when planted. But it should be noted that seeds enclosed and sealed up in these cones are not only surrounded by the natural balsamic elements, but are also attached to the rhachis or axis of the cone, and it is not improbable that they draw some sustenance therefrom during their long imprisonment. It is certain that when removed from the cone their vitality decreases rapidly. When seeds of this kind are to be kept for a year or more they should be left in the cones until wanted for use. Of course the seeds of conifers which naturally drop from the cone when mature cannot be preserved in this way.

Experiments have been, and are still being, made for the purpose or determining the relative duration of life in different kinds of seeds that have been preserved under what are considered favorable conditions, but all show that fresh, new seeds are far preferable to old ones, and these the propagator always endeavors to secure.

Many years ago Professor Alphonse De Candolle, one of the most eminent botanists of Europe, tested the vitality of 368 species of seeds, fifteen years old, all collected in the same garden, and sowed at the same time and under the same conditions as nearly as possible. The following are the results :

Malvaceæ	5 came up out of	10 species.
Leguminosæ	9 " " " "	45 "
Labiatæ	1 " " " "	30 "
Scrophulariaceæ	0 " " " "	10 "
Umbelliferæ	0 " " " "	10 "
Caryophyllaceæ	0 " " " "	16 "
Gramineæ	0 " " " "	32 "
Cruciferæ	0 " " " "	34 "
Compositæ	0 " " " "	45 "

It will be seen by the foregoing list that seeds of the *Malvaceæ* or Mallow family retained their vitality the longest, as one-half of the species germinated after fifteen years, while of the Leguminosæ only one in five germinated, while with barely one exception all of the others failed. Prof. Balfour, in referring to the experiments of Prof. De Candolle, says : "Large seeds were found to retain their germinating powers longer than small ones, and the presence or absence of separate albumen or perisperm did not seem to make any difference. Compositæ and Umbelliferæ lost their germinating power very early."

From these experiments Prof. De Candolle concludes that duration of vitality is frequently in an inverse proportion to the rapidity of the germination.

PRESERVATION OF SEEDS.—While the propagator of plants will usually endeavor to sow seeds as soon as practicable after they are ripe, still it is often necessary, as well as desirable, to preserve them in good condition for a few months, and sometimes for several years. The most favorable conditions for preserving the germinating power of seeds no doubt are such as may be said to accord with natural laws, but not by strict natural methods. Cultivated plants are mostly far removed from their native habitats, and are also subjected to the influence of artificial surroundings; therefore we could not, if we would, adopt nature's methods of propagation ; besides, we aim to improve upon nature, and make a far greater number of seeds grow than is possible in the absence of man's assistance. There can be no general rule given for the preservation of seeds, but by arranging plants into groups we may devise a set of rules which will assist, if they do not prove to be an infallible guide. As I have already stated, the seeds of conifers keep better in the cones, whenever this is practicable, than when removed, and the same rule holds good with many other kinds of seeds that are not enclosed in cones. Indian

Corn will keep sound on the cob much longer than if shelled when first gathered, and the same is true of the Millets, Sorghums, and even of the smaller grasses, if they are stored in a dry, cool place. Seeds of some of the Leguminosæ will remain sound for many years if kept enclosed and sealed up in the pods, and it is well known to seed-growers that the seed of the common Onion will retain its vitality much longer in the heads than if threshed out as soon as ripe. For all kinds of seeds which will admit of any drying, like those of our common vegetables and cereals, there is probably no better method of preservation than to store in boxes or bins, and small lots in cloth or paper bags, and place them in a dry, cool room.

Seeds with hard shells, like the nuts, which require softening or opening of the pores of the shell to admit moisture to the kernel, should be placed in a position where these essential conditions will be assured. In cold climates, frost and moisture will expand the shell, and in warm ones heat and moisture perform the same service. But whether the seeds are to be kept dry or moist, in a high or low temperature, these conditions should be as uniform as possible, extremes of every nature being more or less injurious, even if they do not entirely destroy vitality.

GERMINATION OF SEEDS.—Heat, moisture and air are the principal requisites for the germination of seeds. Light is not essential, and on some kinds of seeds it appears to be detrimental, retarding germination, presumably from its known action in the decomposition of carbonic acid. The temperature required is exceedingly variable, for with the seeds of some tropical plants a hundred or more degrees Fahrenheit are necessary to cause germination, while there are those, natives of cool climates, that will sprout at a temperature of thirty-four or five, or two or three degrees above the freezing point.

As the heat and moisture come in contact with seeds the materials of which they are composed swell and soften, chemical changes follow, rendering the stored up matter fit for nourishing the embryo. In albuminous seeds the starch is changed into dextrine, thence to sugar, through what may be termed the result of contact and the rearrangement of the molecules of the seed. Oxygen is absorbed and heat generated, as may be seen on an extended scale in the operation of malting Barley and other kinds of grain. In exalbuminous seeds slightly different chemical changes occur, but all tend to the preparation of nutriment for the embryo plant. It is quite evident that the chemical changes that take place in sprouting seeds differ as widely as do their chemical properties, but all are set in action by the stimulus (heat) in the presence of moisture.

The increase of heat accelerates germination, provided it is not carried so far as to prevent the natural chemical processes. A temperature of sixty-five to seventy-five degrees may be considered a safe one for most kinds of fruit, flower and vegetable seeds, but those with hard shells or coverings, and especially those of tropical origin, will usually require a higher temperature. In a lower temperature, or less than fifty degrees, the necessary chemical changes proceed very slowly, if at all, and often cease altogether, even after having once commenced, and when this occurs the seed usually decays, and for this reason haste in sowing seeds in spring, and before the ground is warmed by the sun and showers, often gives unsatisfactory results. The germination of seeds is governed by the same principles as that of the production of buds from tubers, bulbs and even the emission of roots from cuttings of ligneous plants, the starchy matter stored up in the cells undergoing very similar chemical changes in the reorganization and growth of the new cells.

CHAPTER VII.

CIRCULATION OF SAP.

Plants obtain the principal part of their nourishment from the liquids and gases absorbed by their roots. The fluids and gases thus absorbed is called crude sap, and this, meeting previously assimilated matter in the cells, mingles with it, and going forward or upward until it reaches the buds, twigs, or expanded leaves, is there exposed to or meets both air and light, producing chemical changes resulting in what is termed organizable matter.

The movement of fluids in endogenous plants is not so readily determined as in the exogenous, owing to the intermingling of the woody and vascular bundles. It is, however, quite probable that both take part in the movement, and as we find cambium near the vascular bundles, it may serve the same purpose as this material in the exogens. But experiments are wanting to show how the transmission of sap takes place in the various and complex structure of endogenous stems; still it is known that there is both an upward and downward flow, but its movement has not been so accurately determined as in the exogenous stems.

The crude sap, or liquid taken in by the roots by the process of imbibition, does not pass upward through open tube-like vessels, but from cell to cell by an endosmose and exosmose action, as explained in Chapter I.; consequently, the crude liquid does not remain separate from the old or previously assimilated sap in the cells, but the new and thinner liquid lessens the density of the older, and both, thus mingled, flow on upward or outward, as the case may be, to the ends of the branches, the result of some force not fully understood. Physiologists do not agree in regard to the cause of motion in the liquids of plants. Some attribute it to what they term capillary attraction, and that the continuous upward flow is sustained through constant evaporation and transpiration which takes place in the buds, leaves and young parts of the plant.

Prof. J. W. Draper attributes the movement of sap to capillary attraction, which he considers an electrical phenomenon. Prof. Leibig takes a somewhat similar view of the phenomenon, and thinks that as evaporation and transpiration take place in the leaves and buds, a portion of the fluids are thus removed and capillary attraction is promoted. Prof. Balfour is inclined to attribute the movement to capillarity in the vessels of the higher plants, and through the process of endosmose the continued imbibition and movement of fluids is chiefly carried on. These movements, he says, will of course take place with greater vigor and rapidity, according to the activity of the processes going on in the leaves, which thus tends to keep up the circulation. Still, if a small or large root of a Maple is severed twenty or thirty feet from the main stem in spring, before the leaves expand, the sap will flow from the wound with as much force as it will from a branch or twig of the same size and the same distance from the base of the stem, a fact that does not appear to establish the theory of capillary attraction.

It is quite evident, however, from what we do know about the movement of fluids in plants, that there are different forces that act and assist in their movements, and it may be due in part to vital force—variation in temperature, or those changes which result from the action of light and air—and partly from capillary attraction following the continuous loss by evaporation, which must constantly affect the density of the fluids, thereby promoting endosmose and exosmose action.

In many herbaceous and acquatic plants there is a rotary or spiral motion of the fluids within each individual cell that can be readily seen with a magnifier of moderate power; and, furthermore, this rotation is constantly in one direction, and if checked and then set in motion again it proceeds in its original course, just as certain twining plants will turn only in one direction. But the rotary motion of the fluids in the cell does not prevent a portion from passing through the cell walls, and the peculiar action is kept up in all so long as active growth proceeds.

Boucherie, in his investigations upon trees in France, found that felled trees continued to imbibe moisture through their exposed cells with considerable force, and that a Poplar ninety-two feet high absorbed in six days nearly sixty-six gallons of pyrolignite of iron. We all know that cut stems of plants, if placed in water, will keep fresh a much longer time than if the lower ends are not immersed, or in some other manner supplied with liquids, and this is mainly, but not wholly, due to the absorption through the exposed cells. It is evident that heat and light have a powerful influence in the flow of sap in plants, by promoting transpiration and action in the cells, but imbibition of liquids by the roots does not necessarily cease with growth of the plant, or even loss of foliage, for as liquids of less density than those within them are presented to the roots, absorption must continue, although the movement may be slow when the plant is less active than during the growing season. We conclude that this must occur from the fact that trees, shrubs and other plants, while apparently at rest, even in cold climates, become gorged with liquids, and at a season when there cannot be any considerable exhalation from the leaves of evergreens, or the twigs and buds of deciduous kinds, which would promote or cause continued absorption of liquids by the roots; still, it is well known to every investigator that exhalation from the parts of plants exposed to the air does not cease altogether, even in the coldest weather, and the loss of this moisture must be made good from the parts below. Whenever there is a total cessation of the movement of fluids through the stems and branches death follows, and in ligneous plants the wood, bark and buds become dry and shriveled. We may, therefore, conclude that the entire sap of trees never becomes frozen solid, and that there is always a flow of gaseous matter, if not of heavier liquids, through the cells, even when the plants are in a semi-dormant state. The often repeated experiment of forcing into growth under glass a cane of a Grapevine or branch of a fruit tree while attached to the parent plant, remaining out of doors and apparently frozen, shows that there must be some communication between the semi-dormant parts and those within the house.

The first effect of light and warmth in spring is to stimulate action in the plants. The fluids absorbed from the soil by the roots are carried upward from cell to cell, through the alburnum or sap-wood of exogenous stems, to the leaves and buds, where they are exposed to air and light, and there changed into organizable matter through a process which is termed assimilation. Some of the liquid part of the sap is exhaled, passing off into the atmosphere, while a portion of the assimilated matter goes to aid in the prolongation of the twigs, enlargement of the leaves, buds, flowers and fruit, and other portions are spread over the entire surface of the plant through the liber or inner bark, even extending down to the extremities of the roots, adding to their size and prolongation. In this way the concentric layers of wood are formed on the outside of the stem and branches. If there is no cessation of growth during the summer the newly-formed cells coalesce, producing a homogeneous, uniform concentric layer which may be readily distinguished from those of previous years. This is the usual method of growth in exogens, but the sap may be diverted from its natural course, for if obstructed, the tissues will change their functions and propel the fluids in other directions through the cells instead of lengthwise with the grain of the wood, as may be readily proved by removing alternate sections of wood from the stem of a tree. Prof. Lindley, in referring to the functions of the alburnum and liber in trees, says: "The two have equally important offices to perform; the alburnum giving strength and solidity to the stem and conveying sap upwards; the liber not only conveying sap downward, but covering over the alburnum, protecting it from the air and enabling it to form without interruption. It is therefore indispensable to the healthy condition of plants that neither the alburnum nor liber should be injured."

The inner layers, or heart-wood, of trees are dead, and they may be removed entirely without serious injury to the living parts, as often occurs, and as seen in hollow trees, which sometimes live for centuries in this condition, new layers of alburnum being annually added to the outside. It is now quite generally conceded that the annual increase in the diameter of exogenous stems is due to the multiplication of the cells of the cambium layer, and the material from which they are formed—or at least the greater part of it—descends in the bark; but there have been, and still are, vegetable physiologists who deny the existence of any distinct downward flow of organizable matter through the liber. Dr. J. M. Schleiden, in his "Principles of Botany," emphatically denies any such movement in plants, and says: "As water is continually exhaled by plants in proportion to the dryness, motion and warmth of the air, so the sap becomes concentrated, and thus interrupts the endosmatic process toward the other cells; this action is continued naturally downward toward the roots, by which new watery and unassimilated fluids are absorbed. If this stream of crude sap is artificially interrupted in its course from below upward, the sap in the upper part becomes more concentrated, and its organizing power increased. This is the simple fact which lies at the foundation of all those phenomena which are brought forward to support the groundless hypothesis of a decending bark sap." In another place he says: "When an Apricot graft grows from the trunk of a Plum tree the latter is naturally and by degrees clothed with Apricot wood, for out of the same soil an Apricot tree would merely take up the same crude material as the Plum tree," etc. But those who oppose this idea of a downward flow of organizable sap in the liber appear to have overlooked the individuality of the functions of the cells, and the fact that while one set or group may be secreting one kind of substance, or performing certain functions, another group may be doing something quite different, as I have already explained elsewhere.

Practical propagators of plants know that the cells of the stock and those of the cion always remain distinct in each, preserving their individual type, and even the old and excretory bark enclosing them retaining its peculiar original characteristic. Pears grafted on Quince roots never change the latter into the former, and we may build up a tree of alternate sections of Pear, Quince, Thorn and Medlar, and each section or part of the stem will retain its individuality, although the roots may be of one species and the branches and leaves of another. The cells of each take from the passing crude sap, or descending organizable matter, the materials needed to build up and retain their own individual structure.

If there were no descending organizable sap in the liber or inner bark, then the girdling of trees would have no more than a temporary effect upon growth; but the pioneers in our American forests have proved to us that to remove a ring of bark and the mere severing of the outer layer of alburnum will cause the death of any kind of tree within a twelvemonth. By this simple process the upward flow of the crude sap is not prevented, for it passes freely through the unsevered alburnum layers, reaching the leaves as usual, where it is assimilated; but the descent of the organizable matter can pass no further than the annular incision in the stem; the result is that none reaches the roots, the latter perish in conse-

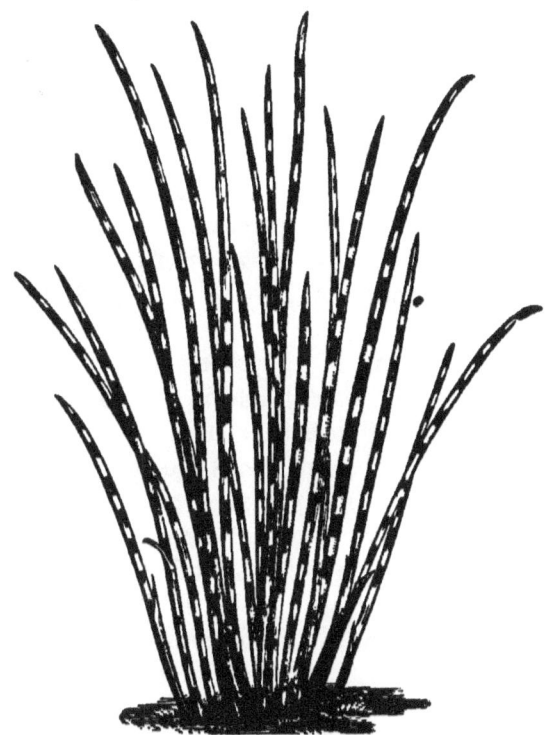

Fig. 42.—VARIGATED RUSH.

quence, and the following season the tree is dead—not a bud on either root or branch survives. Those species which are the most obstinate in producing suckers from their roots are readily killed out by so simple an operation as girdling in late winter or very early spring.

Ringing, or the removal of a ring-like piece of bark from a single branch of a fruit tree, or cane of a Grapevine, has long been practised for the purpose of producing extra large specimens of fruit, it becoming gorged with organizable matter, which is prevented from passing downward on account of the removal of the ring of bark.

The power of cells to appropriate certain elements and not others can scarcely be doubted, although we know but little of the process, but it is through the individuality of their functions that they are enabled to take from the liquids passing through them those materials necessary for their own growth, allowing others to pass unappropriated; thus certain groups of cells assume one form, size, color, as in the case of the double or quadruple grafted Pear tree, each group retaining its own individuality to the end. This preservation of individual characteristics of cells may not only be seen in the multiple stems of grafted trees, but frequently in various simple herbaceous plants, as in the stems and leaves of the Japan Zebra Grass (*Eulalia*), and in the Variegated Rush (*Scirpus*), shown in figure 42. In both of these plants the long, slender leaves and stems are composed of alternate sections of white and green parts, and if we take no account of any downward flow of organizable matter, and presume that the crude sap is assimilated in its progress upward through the leaves, we still find it difficult to explain how the green colored liquids can pass through the white parts and leave no stain of chlorophyll, except upon the hypothesis that there is an individuality of action in the various groups of cells of the leaves which enables them to retain their distinct characteristics, while all are drawing nutrients from the same source and through the same set of vessels. We know that some such action takes place in grafted trees with stems and roots made up of different species. Quince roots may imbibe nutrients from the soil, while Pear leaves may assimilate them; yet each remain true to its species or variety. Conductibility in this case is not only a function of the cells, but in the operation a kind of selection or choice is made of materials passing through them; but this elective power appears only to extend to those elements which are of a congenial nature to the whole or some part of the plant, for when certain poisons are presented to the roots or leaves they are unable to resist them. This act of non-resistance does not prove that plants have no power of selection of nutrients, but merely that they are incapable of resisting certain elements which are inimical to their health and growth. The difference in the density of liquid poisons presented to the roots may also have some influence upon the elective powers; and it is well known that some mineral poisons, when much diluted, will be absorbed by plants without any apparent injury, while stronger solutions will cause death, as they also do when given to animals in large doses. Certain gases are also highly injurious to plants, being readily absorbed by both roots and leaves. Sulphuretted hydrogen gas, says Dr. Balfour, attacks the leaves at the tips first, gradually extending to the leaf-stalk, which would seem to show that it followed the return flow of sap. Sulphurous acid gas is highly injurious to plants, as many a gardener has learned to his cost through the accidental ignition of sulphur in houses filled with plants. But the gas which most interests the practical cultivator of plants is the one known as carbonic acid gas, for it is from this that the carbon, the most abundant single element of plants, is obtained. Carbon also makes up some forty to fifty per cent. of the bulk of the ordinary plants cultivated for food, and in trees the proportion is still greater, all of which is supposed to be derived from carbonic acid gas, but just how it is obtained or fixed in the form of plant cells has been a mooted question among vegetable physiologists and chemists ever since they began to investigate the principles of growth and composition of plants. This carbon ghost will neither remain passive or find an abiding place either in the atmosphere or soil; for while one authority allows it free entrance to the roots, another is equally certain that it only finds its way into the plants through the leaves. In a recent work of Dr. M. T. Masters, in speaking of root action, he says: "No passage of acid fluid out of the cell takes place, root excretions having no existence," etc. But in another paragraph on the same page he says: "On the other hand, roots do not absorb carbonic acid gas nor exhale oxygen as the leaves do (in the sunlight), but they do give off carbonic acid gas, which, with the aid of water, converts the insoluble carbonates of the soil into soluble bicarbonates, and exercise a similar power of solution in the case of phosphates." If the roots "do give off carbonic acid gas," it is certainly by an "excretive" action, which the author utterly denies in the first paragraph quoted. In describing the action and function of leaves he says: "The paramount function of the leaf is the absorption and assimilation of carbon. Carbon, as such, does not exist in the atmosphere, unless, indeed, as an impurity in the air of towns, and is a very prejudicial one to plants. It is in the form of carbonic acid gas—a combination of carbon and oxygen—that it is found in the atmosphere, but only in small proportion compared with the other constituents. In the plant, carbon exists in much larger proportion than any other ingredient, with the sole exception of water. It forms, in fact, fifty per cent. of the dry matter of plants left behind after the water and gases have been expelled by heat. This large quantity of carbon has to be taken up in the form of *carbonic acid by the leaves.*"

There can be no question in regard to the meaning of Dr. Master's words, for they are direct and to the point —namely, the carbon of plants is derived from the air and through the leaves only, consequently from a source where this important element is the least abundant, and, we may say, so largely diluted that our chemists tell us that it averages only about one twenty-five hundredth part of the bulk of the atmosphere, while it abounds in many soils, springs, brooks, wells, caverns in the *humus* of forests, besides occurring combined with bases, forming carbonates of lime, magnesia, soda, strontia, baryta, as well as in the various oxides of metals, some of which are always found in fertile soils. The position which Dr. Masters takes in regard to the source, as well as the way in which carbonic acid is utilized by plants, does not differ essentially from that held by several other European botanists, but he gives the theory in a better form than I have seen it elsewhere.

Dr. Balfour, however, a very eminent English authority on vegetable physiology, says that "carbonic acid is readily taken up, either in its gaseous state by the leaves, or in combination with water by the roots." We certainly can scarcely conceive of any inherent power of choice so potent in the roots as to enable them to reject this gas which is so necessary to their health and growth, and it is well known to be far more abundant in the soil about the roots than in the atmosphere. In a recent (1885) edition of the "Manual of Agriculture," by Geo. B. Emerson, I find the following in relation to this subject: "Carbonic acid is the most indispensable and abundant article of the food of all plants. It enters the plant dissolved in water, and either remains in that state, or the vital action of the plant, in the light of the sun, decomposes the acid and throws back most of the oxygen into the atmosphere, but retains a portion which performs important offices, and also retains the carbon. This forms the solid parts of every plant."

Among men who combine science with practice the foregoing is the most generally accepted theory of the manner in which plants obtain the great bulk of their carbon. It is open to some objection, but it will mislead no one in the management of plants. All practical cultivators of plants recognize the great value of carbonaceous matter, not only as found spread over the surface of the earth in forests, prairies and plains, but in the manures applied to soils from which it has been exhausted by long cultivation or otherwise. It is not denied, however, that plants do absorb carbonic acid gas through their twigs, buds, leaves, and other appendages; in fact, as they take in moisture through these parts, other nutrients must also pass in combination with it, as well as carbonic acid gas, but in infintesimal quantities compared with the amount absorbed by the roots.

In this respect plants do not differ greatly from animals, for it is well known that nutrients and poisons may be conveyed by air to the blood through the lungs, or they may be thrown directly into the system by hypodermic injections through the skin; that a starving man would live longer in an atmosphere laden with the fumes of cooking meat than in one from which such fumes were entirely absent; still, few of us would care to take all our beefsteak in this way.

I have been prompted to refer to this subject of the sources of carbon in plants somewhat at length, because it has of late become quite a habit among writers on agricultural and horticultural topics to ignore the value of carbonaceous matter in the soil, and some go so far as to doubt the value of ammonia and other sources of nitrogen; in fact, they would lead us to believe that with air and water, and a few broken rocks for an anchorage, the husbandman will be enabled to produce the most luxuriant growth in plants of all kinds. But it will be well for the cultivators of plants to continue the practice of applying liberal quantities of carbonaceous and nitrogenous manures to their soils, trusting to these to supply the greater part of the nutrients, and these through the roots of plants, or by placing the food where it will be most likely to be utilized.

Leaves have various functions; they permit of the evaporation of superfluous moisture and gases, as well as the absorption of the same from the air, as conditions are changed. So long as the leaves are supplied with an abundance of moisture through the roots and stems they will take none from the air, but let the supply be cut off from below and have it presented to the leaves, and a reverse action soon follows. We may test this reverse movement very readily with leaves and twigs cut from growing plants, and at the same time it may be seen that the absorption of moisture varies under different circumstances and conditions. The force exercised by the roots is by far the greatest; next in degree is the absorption of liquids by the exposed cells in the severed stem, and last and least through the pores or stomata of the leaves. Of the many experiments I have made in investigating the movements of liquids in plants, I may refer to one which has a direct bearing on this question of absorption of moisture through the stomata of the leaves. I cut off small, tender branches and leaves of various plants, such as *Achyranthus, Coleus, Verbenas, Abutilons*, etc., spread them all out on a shelf in the sun, where they soon began to flag, and after all had become fully wilted, the severed ends of one-half of each variety and kind were dipped in melted wax in order to hermetically seal up the exposed cells. Then all were placed in a box and lightly sprinkled with pure water. At the end of twelve hours they were taken out and carefully examined, and the difference in the condition of the sealed and unsealed was quite apparent; those with exposed severed cells had fully revived, while many of the others were still limp, but all showed signs of recovery. The cuttings were all replaced in the box and left there another twelve hours; then all were found to have recovered and to be as fresh as when first taken from the parent plants. This experiment shows that absorption of moisture may take place through the severed cells of the stems, leaf-stalks, and also through the stomata, under varying conditions and circumstances.

Dr. G. Hartwig, in "Harmonics of Nature," says: "The stomata are destined to admit air, not water, which by drenching the leaf would entirely interrupt the process of respiration." But Hugo von Mohl, a very able investigator in the structure of plants, has shown very clearly that the stomata open in the presence of moisture and close in a dry atmosphere, as may be readily verified by any one who will examine them under a glass of high magnifying power.

Gardeners utilize this power of severed cells to absorb moisture and nutriment, especially in propagating plants by cuttings of the young and immature parts, or by what are technically termed "green cuttings." In employing such cuttings it is well for the propagator to keep in mind the fact that the severed and exposed cells at the base of the cuttings cannot resist the noxious properties which may be presented to them in liquids with the same power as roots, hence the greater necessity of selecting pure and innocuous materials in which to plant the cuttings while producing roots. I have kept the roots of such coarse celled and rank growing plants as the Calla (*Richardia*) submerged in a solution of Madder and other vegetable dyes for weeks at a time without detecting any indications of their absorption by the plants, although the outer bark, or epidermis, of the roots and smaller rootlets were soon colored; but if a leaf-stalk of one of these plants is cut off and the severed end set in the dye, some of the coloring matter will soon be absorbed and easily traced upward to a greater or less distance.

That the imbibtion of liquids through the roots, as well as by the severed cells of a green cutting, is in part due to leaf action can scarcely be questioned, but it is not the only force that aids in the ascent of the sap of plants, for absorption will occur in the absence of leaves and even in the dead tissues of plants, but with less force than in living ones.

In plants that have no distinct leaves, like most of the Cactuses and Stapelias, the epidermis of the stems performs the functions of true leaves in other kinds of plants, but just what these functions are has not as yet been fully determined, although many plausible theories of leaf action have been given by vegetable physiologists and chemists, but as they do not agree we may safely conclude that there is not only some mystery surrounding this matter, but that there is still something to be learned about leaf action as well as the chemical changes which take place during the process of assimilation. Plants, like animals, to a certain degree possess an inherent power of adapting themselves to varying conditions, not being controlled by such invariable laws that their lives are jeopardized by every change of temperature, hygrometrical condition of the earth and air, or variations of light and shade. It can be readily demonstrated that sunlight is necessary for the production of chlorophyll, or the green coloring matter in the leaves, but it is no more so than for the depositing or production of other natural colors of the leaves, and there are no good reasons for supposing that the green matter in the leaf of the common Beech tree is of any more importance or obtained from a different source than the red pigment in the leaf of the Purple Beech, or the colors in the leaves of *Coleus Verschaffeltii*. Neither will it be produced in the absence of light; still, the deepest green in the foliage of plants in general is not to be found in countries where there is the most sunlight, but in those where there is alternate light and shade, with a decided preponderance of the latter. In the deep shade of our forests the Kalmias, Rhododendrons and Hollies not only thrive best, but it is in such situations their leaves assume the deepest shade of green. Alternate sunlight and shade, with abundant moisture, are the requisites for coloring the leaves of plants a deep green, and in Ireland, long known as the "Emerald Isle," they do not have half the sunlight we do, while under the almost cloudless skies on our Western plains the foliage of plants have a sickly yellowish or grayish green.

CHAPTER VIII.

SEX AND FERTILIZATION.

At what stage in the evolution of plants differentiation of sex becomes a distinctive characteristic, has not been fully determined. But from what we know of their development, it is quite evident that distinct sexual organs are the result of a progressive movement from the lower to the higher and more complex organisms. Nothing like sexual organs have been discovered in the simple one-celled plants, or even among those much higher in the scale, like the Mushrooms, Mosses and Lichens, and even in the Ferns and other cryptogamous plants the sexual organs are not clearly defined, although in some they are sufficiently distinct to be utilized in what is called cross-fertilization or hybridizing of species. It is quite probable that in the lower forms of plants the conjugation of the sexes occurs by a simple coalescence of cells, somewhat as two drops of water brought in contact unite and become one. But as the practical propagator of plants will seldom have occasion to investigate the sexes of the lower orders, they may be passed over here without further attention.

In a large majority of the plants under cultivation the sexual organs are sufficiently distinct and conspicuous to be readily examined and manipulated, whenever there is an occasion to interfere with the natural processes of reproduction. In a state of nature, the sexes are generally placed in positions favorable to direct intercourse, either by contact, or through natural vehicles, for the transmission of pollen from the male to the female organs. If the sexes are widely separated on different plants, or on different parts of the same plant, wind and insects, either or both, become the media for transporting pollen, and every person who is at all observing, and takes an interest in such matters, must not only have noticed the clouds of yellow dust blown from Pine, Chestnut, and similar kinds of trees, but also the pollen-laden bees and other insects that pass from flower to flower, in search of honey or whatever may serve them for food and other purposes.

While in the larger majority of plants both anthers and pollen-grains are of a yellow or yellowish-white color, still there are many exceptions, and red, brown, blue and other shades of color are seen in the pollen even among the plants of the same family or genus.

In some plants there appears to be considerable mechanical force required for the proper distribution of the pollen; this is exhibited in a peculiar manner in the native Kalmias. In these plants the stamens are so arranged that they are bent back with the expansion of the flower, and held in this position for several days, and when relieved by the petals they spring back to the center, striking the stigma with considerable force, the anther cells bursting at the same time and widely scattering the pollen. The position and form of the stamens in the flower of the Kalmia are shown, greatly magnified, in figure 43. In some the female organs are shorter and placed below the male, the pollen dropping from the anthers upon the stigmas; or in drooping flowers, like the Fuchsia, the pistils may be many times longer than the stamens, but if extending below them they receive the pollen as it falls from the anthers above. There are almost as many different ways in which the ovule is fertilized as there are different plants. The stamens and pistils are also of different size and form, and in some plants, as in the Conifers, Palms, etc., the pistils are entirely wanting, the pollen coming in direct contact with the exposed ovule. But however variable the sexes or form of the sexual organs in the higher orders of plants, their functions are very nearly identical, and in performing the operation of artificial fertilization we proceed in about the same manner with all, merely interfer-

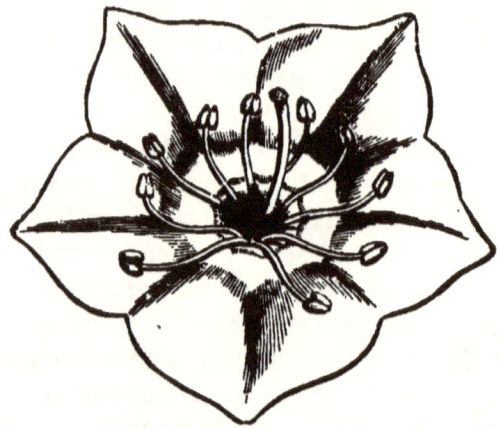

Fig. 43.—FLOWER OF KALMIA LATIFOLIA, ENLARGED.

ing with nature sufficiently to produce varying results. If both male and female organs are present in the same flower, we remove the former in order to prevent them from performing the operation for which they were intended by nature, and then introduce pollen from some other closely allied plant to fertilize the ovules. But where the sexes are in separate flowers on the same plant or on different plants, we have only to protect the female organs against the visits of insects, or contact of pollen transported by the wind, and then introduce that which we desire should perform the act of fertilization. For

SEX AND FERTILIZATION. 103

instance, if we wish to produce a cross or hybrid plant between the common White Lily (*Iilium candidum*),

Fig. 44.—LILIUM CANDIDUM.

figure 44, and some other species or variety, we watch for the opening of the flowers and expansion of the petals, *A*; then we cut off the anthers, *B*, to prevent self-fertilization through the pollen from them falling upon the female organ or stigma, *C*. We then procure pollen from the flowers of the other variety or species, and apply it to the stigma of the first, or White Lily. The stigma, when in the proper condition for fertilization, is coated with a viscid substance to which the pollen grains will adhere quite firmly, consequently there is little danger of their removal while handling the flower. To prevent insects from visiting the flowers, and interfering with our operations, the flowers should be enclosed in bags made of muslin, mosquito netting, or some similar material soon after opening, and kept enclosed until fertilization is assured; then removed to relieve them from further restraint. It is always advisable to repeat the application of pollen, because the first may not "take," owing to the immaturity of the stigma. Pollen, however, is not such a delicate material that its potency is readily destroyed by rough handling, and that produced by some kinds of plants may be preserved in good condition for use several days, or even weeks, if excluded from the air. This fact should be kept in mind, for it often occurs that the plant from which we desire to obtain pollen blooms in advance of the one on which we desire to use it.

In the application of pollen a small camel's hair pencil is the most convenient implement, but where the anthers are large, as in the Lily, a pair of forceps, or even the fingers, may be used for transferring it from one plant to the stigma of another.

The pollen grains placed on the stigma burst open and discharge their contents upon it, where they may be said to germinate, sending down long flexible filaments or "pollen tubes" through the loose tissues of the stigma and style to the ovary, or true seed-vessels, where the completion of the act of fertilization takes place. The exact nature of this act is not fully known, but the mechanical part of the operation is as stated, as may be determined by dissecting the stigma and style at various stages of the growth of the pollen tubes.

Prof. Sachs, in referring to the fertilization of seeds, says: "Since every ovule requires one pollen tube for its fertilization, the number of tubes which enter the ovary depends, speaking generally, on the number of the ovules contained in it; the number of pollen tubes is, however, usually larger than that of the ovules; where these latter are very numerous, the number of pollen tubes is therefore also very large, as in the Orchideæ, where they may be detected in the ovary, even by the naked eye, as a shining, white, silky bundle."

While it is no doubt true that each ovule must—except in rare instances—come in contact with a pollen tube in order to be fertilized, it is not at all necessary or probable that a distinct and separate tube or filament starts from the pollen on the stigma, thence passing through the style to each ovule to be fertilized, especially in those plants where there are many seeds produced in a seed-vessel, as in the Lilies, Mallows, etc. But instead of the multiplicity of pollen tubes that would be required if one was formed for each ovule, several ovules become attached to the side of one pollen tube as it passes lengthwise through the carpels. As the seed grows or enlarges the little branch of the tube can be readily seen. Where the pistils correspond in number with the cells in the seed-pod, as in the *Abutilons*, one pollen tube suffices for the fertilization of the two to six or more ovules in each of the eight to twelve loculicidal cells of our common cultivated varieties. In some of the Malvaceæ, there are a greater number of pistils, or at least branches of the style and stigma, than there are cells of the ovary or even ovules; consequently, if each stigma produced a pollen tube, they must either coalesce in their growth or some of them become abortive.

In such plants as the Strawberry, Rose and Indian

Corn, and closely allied plants, there is one pistil for each ovule or seed-vessel, and more than one pollen tube would be an entirely superfluous production, hence we find only one in each. If a pistil is destroyed the ovule at its base remains unfertilized, and no grain or seed is produced, and where a few of the pistils are fertilized and the others not, the result on an ear of Corn will resemble the one shown in figure 45, the grains on it varying in number with the number of pistils fertilized. It is only about sixty years (1823) since Prof. Amici, an Italian botanist, discovered the pollen tubes, and this opened a new field for investigation, which was soon occupied by some of the most eminent botanists of Europe. Previous to the discovery of Amici, the process by which the ovules were fertilized was unknown. Some vegetable physiologists supposed that the pollen grains passed bodily through the pistils to the ovary—an erroneous idea which still prevails among certain horticulturists of the present day.

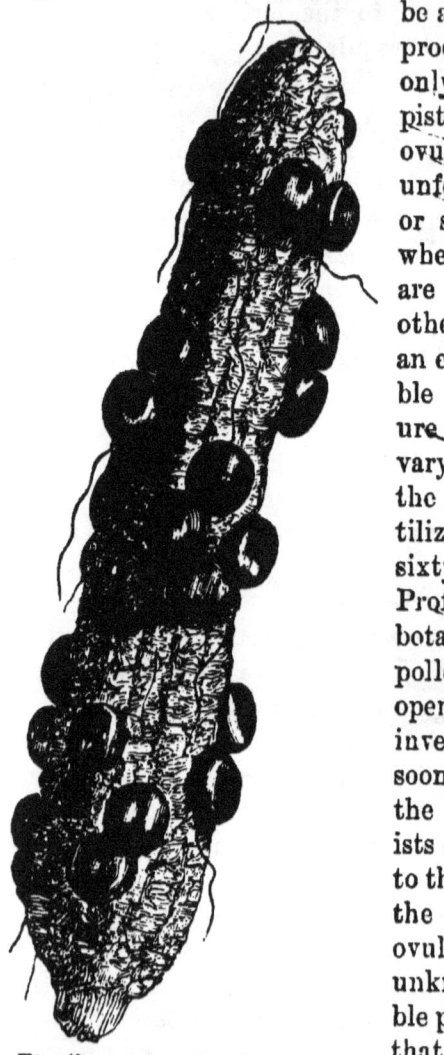

Fig. 45.—IMPERFECTLY FERTILIZED EAR OF CORN.

When the ovule of one plant is fertilized by the pollen of another, the seed resulting therefrom should, as a natural result, produce a plant intermediate between the male and female parent, but it is seldom that both parents possess equal potency in transmitting their own individual characteristics to their offspring. For instance, we might be successful in fertilizing the ovules of the White Lily with pollen taken from one of the many yellow or red varieties in cultivation, and yet the plants raised from the cross-bred seed may all resemble the female parent more than the male. This alone would not prove that the artificial fertilization had failed, but merely shows that the influence of the male parent was less potent in transmitting its characteristics to the offspring than the female. A second generation of seedlings from the cross-bred or hybrid plants may show more of the characteristics of the staminate than the pistillate, or the reverse of those of the first generation. It may also be well to bear in mind that in experimenting with wild plants, or those recently introduced into cultivation, we have to contend with inherited characteristics acquired by close interbreeding through an unknown number of generations, and these have become so fixed in vegetable structure that introduced disturbing causes have, at first, but a slight influence in producing any very pronounced change in form or structure of offspring. Still, as a rule we expect the offspring of cross-fertilized seed will show the effect by varying more or less from the parent type, and when once a species commences to vary, as a result of artificial causes, it becomes quite difficult to determine the limits.

In all of our operations in transferring the pollen of one plant to the stigma of another, we proceed in very much the same way as in the cross-fertilization of the Lily, only varying the operation to correspond with the variation in the structure of the flowers of different kinds.

In some instances it may be necessary to remove a part, or force open the petals or other organs that envelop those to be operated upon. It is also advisable, in many instances, to anticipate nature in such operations by a few hours, or even days, and place the pollen in a position where it will be utilized by the stigma when required.

In crossing and hybridizing the Grape, it is a common practice to anticipate nature by several hours, because its flowers do not expand like those of the Lily, Rose and similar plants, but instead, the petals cohere to one another at the top, breaking loose at the base, and are then forced upward by the elongation of the stamens and pistils, as shown in figure 46, *A*, the petals being thrown

Fig. 46.—FLOWERS OF THE GRAPE.

off in the form of a cap. The five stamens then expand, as shown in figure 46, *B*, these surrounding the pistil, *C*. The anthers should be immediately removed with a pair of small and sharp-pointed scissors, leaving them as seen in the figure at *D*. In the illustration, figure 46, the Grape flowers are shown somewhat enlarged. Before commencing to operate upon an immature cluster, it is well to thin out the undeveloped flowers, not only to facilitate manipulation, but also to prevent the crowding of the fruit when fully grown. An immature cluster of flowers thus prepared must be closely examined from day to day, and so soon as the petals break loose from their base they should be carefully thrown off with the point of a knife or other sharp-pointed implement, and the anthers removed as directed. Pollen may be immediately placed on the stigma or the operation deferred for a few hours, but it is better to apply it as soon as the cap is removed, as the minute grains will usually adhere and be in position for absorption, or, more properly, germination, when required. As the flowers do not all open at the same time the cluster may have to be operated upon for several days in succession, and when applying pollen to those recently opened it is well to touch the stigma of those operated upon the previous day with fresh pollen, in order to increase the chances of success. Repeat these operations until all the flowers, or as many as may be required, are fertilized, and then remove all that remain unopened. The cluster should be protected by enclosing it in a bag made of some kind of thin fabric like Swiss muslin or mosquito netting, for if some such covering is not used, insects may visit the flowers, and interfere with our work. When the Grapes are ripe they should be gathered, the seeds removed and planted in the usual manner. In operating on the Grape as described, it is presumed that the flowers are perfect, containing both stamens and pistils as are usually found in the common wild and cultivated varieties. But among the species of the Grape indigenous to North America, an occasional plant is found bearing both perfect and imperfect flowers, while others produce only staminate flowers, the pistils being undeveloped or deformed. When perfect and im-

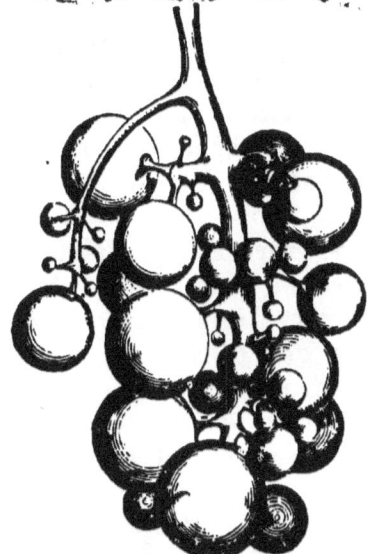

Fig. 47.
IMPERFECT BUNCH OF GRAPES.

perfect flowers exist in the same cluster, the bunches of fruit will be equally irregular, as shown in figure 47, from the "Grape Culturist," page 14. But on vines bearing only staminate flowers, or those with fully developed stamens and deformed pistils, no fruit is possible or is ever produced. There are many vines of this kind to be found in different parts of the country, and some very old ones have been preserved more as a curiosity than for intrinsic value. In the Spring of 1860 I saw one of these staminate vines in the grounds of T. S. Kennedy, Esq., Louisville, Ky., which was then supposed to be over seventy-five years old, and although it bloomed freely almost every spring, it had never been known to produce a fruit, its flowers, like others of this sex, exhaling a fragrance somewhat like that of the Mignonette.

But while occasional abnormal forms are to be found among wild plants of all classes, orders and genera, they may be considered as exceptions to a general rule, while among cultivated plants it is almost the reverse, for suppressed, deformed or enormously developed sexual organs and malformations of the various parts and appendages are to be met with almost everywhere. In a strict botanical sense, all such variations from normal types are monstrosities, and yet double flowers, seedless fruits, misshaped and discolored foliage are greatly prized and usually considered in the light of valuable acquisitions. Owing to this wide departure from normal types, as seen in all classes of cultivated plants, it would require far too much of the space at my command to give full and definite directions for crossing and hybridizing the vast number of different species of plants belonging to the various classes, or even the members of some of the larger families. But I may remark, in a general way, that when a person possesses the inclination to perform such operations, he will naturally begin to observe the form, structure and habits of plants, and soon, with the aid of some

SEX AND FERTILIZATION.

elementary or advanced botanical treatise, be able to distinguish the sexual and other organs and parts of flowers, and when a person has acquired this knowledge through actual observation and study, that which may have been previously obscure will, in a great measure, become plain and easily understood. Partly or wholly smothered organs will be relieved by removing, entire or in part, others that have overgrown and shaded them, as is frequently practised in removing the abnormal and highly-developed petals of double flowers, as in the double Dahlias, Chrysanthemums, Asters, and other plants of the Composite Family.

In the doubling of such flowers as Fuchsias, Carnations, Camellias, Roses, and all members of the great Rose Family; Apple, Peach, Plum, Cherry, Almond and Quince, as well as those of the Mallow Family; Abutilon, Hollyhock, etc., the additional number of petals are mainly transformed stamens, and the metamorphosis of these organs can be readily traced in their gradual advance from the single to the double form. There are, however, exceptions to this rule, and the multiplication of petals is a distinct process from that in which they proceed from transformed stamens and pistils. Sometimes we find a duplication of the petals in the Hollyhock, Rose of Sharon and Chinese Hibiscus, while the sexual organs retain their normal number and form. In the only double Abutilon at present known (*Abutilon Thompsonii pleno*), the stamens are all transformed into irregularly shaped petals, with no duplication of the divisions of the original corolla; but a plant of *Abutilon,* "*Mary Miller,*" in my greenhouse, recently produced a flower with a perfectly duplicated corolla, or a semi-double flower, showing that what we call "doubling" may proceed in this genus from both multiplication of the corolla and the transformation of the stamens.

In some plants, like the Rose, Fuchsias and Abutilon,

the pistils in the center of the flower are the last to be effected by cultivation, and often remain in a condition to perform their natural functions long after the stamens have changed to petals, and with a little assistance, to prevent smothering the stigma, may be readily fertilized artificially, and fertile seeds produced from quite double flowers. There is, however, a limit to all operations of this kind, as well as to our knowledge of vegetable structures, as, for instance, we occasionally find plants which appear to have perfect sexual organs, and yet they resist all efforts to make them fruitful, but why this is so we are unable to determine.

LIMITS OF CROSS-FERTILIZATION. — The limits of artificial fertilization of plants have never been determined, and they only can be through the aid of innumerable and oft-repeated experiments, and if we could decide what is possible with plants, as they exist at this time, new forms must necessarily appear as the result of artificial intermingling of species, thereby opening new and at present unknown fields for experiments and investigations. In the ever-changing phases of plant life, who can say that the impossible of to-day will not be possible to-morrow or a few years hence?

Under ordinary circumstances, varieties of a species may be cross-fertilized far more readily than species can be hybridized. The distinction between the offspring of species and varieties is not so generally recognized as it should be among cultivators of plants. Correctly speaking, a hybrid is the offspring of two species. For instance, if we should take the native Apple of Europe (*Pirus Malus*), which is the parent of nearly all of our cultivated varieties, and the American Crab Apple (*P. coronaria*), and by fertilizing the flowers of one species with pollen from the other, produce a plant with the characteristics of both combined, we would then have a proper or true hybrid. But if we fertilize the flowers of the Baldwin Apple with pollen from the Porter, or any other variety of the same species, the offspring will only be a cross-bred variety, and in this kind of crossing we only intermingle elements that may have been intermingled many times before.

True hybrids may be considered as forced productions rarely found in nature, and the few that have been produced without the assistance of man, are but exceptions to the rule. At one time, and that not many years ago, hybrids among cultivated plants were so rare that it was thought they must necessarily be barren, or nearly so, as was supposed to be the case among hybrid animals—the common mule or offspring of the ass and mare, being the accepted type of such animals; but not only has the mule been known to breed, but many of the hybrid plants are as productive as either of their parents. Unfortunately, however, for the student of natural history, it is frequently very difficult to determine species from varieties among both plants and animals. Every collector in any branch of natural history, who has attempted to arrange his specimens in the order laid down in "check lists," or the works of the highest authorities, knows, to his cost, how frequently he is compelled to re-arrange his cabinet to meet the ever-changing opinions and discoveries of those to whom he has looked as competent guides in such matters. If he seeks specific information that will enable him to determine what is or is not a true species, he will find but little that is clear and definite on this point. Prof. Asa Gray says that the "idea" of species is "based upon a succession of individuals, each deriving its existence, with all its peculiarities, from a similar antecedent one, and transmitting its form and other peculiarities essentially unchanged from generation to generation. By species we mean absolutely the *type* or original of each sort of plant, or animal, thus reproduced by a perennial succession of like individuals, or, concretely, the species is the sum of such individuals." But as we know little or nothing of the "type or original" of what we now call a species, it is very difficult and often impossible to distinguish them from varieties; or, in other words, where there are various closely allied normal or wild varieties, each extending over extensive areas of country, or even when they are more or less intermingled, it is scarcely possible to determine which is the original type or species. Dr. W. B. Carpenter, in referring to this subject and the tendency of some species to run into spontaneous variations, for which no external cause will account, very truly says: "Hence, in discriminating what are real species from what are simple varieties, the botanist is treading on very insecure ground, until he has ascertained, for every species, its tendency to run into varieties of form, whether spontaneous or induced by change of external conditions. His greatest difficulty arises from those cases in which have arisen what are termed permanent varieties, which reproduce themselves with the same regularity as do real species." It may be said on this subject that the most thorough and experienced investigators are the least positive in determining what should or should not be called a species, while the superficial writers and observers are usually quite ready at all times to decide such questions to their own satisfaction, if not to that of any one else.

If a plant in its wild or cultivated state reproduces itself from seed with slight or no variation, this fact would not prove it to be a distinct species, but merely show that the natural forces of the plant were very nearly or perfectly balanced. It may be said, in a general way, that species differ from varieties by possessing characters that through a longer period of interbreeding have became more firmly established under uniform conditions.

Plants in a state of nature perpetuate their species and

varieties with great uniformity, but when removed from their natural habitats, a change in their offspring is usually observable, and these variations may be intensified by cultivation and other external causes. The first variation may be considered the entering wedge, which will, if followed up, divide and disintegrate the most stubborn of species.

The novice in horticultural matters will necessarily need some other source of information in regard to species and natural varieties than that acquired through his own personal observation, and this is found in our standard botanical works, but it may be well to bear in mind that their authors are as other men, not wholly infallible, but probably as near right as the present state of botanical science permits; also that the acquirement of true knowledge often tends to change the opinions of the most learned.

Taking our standard botanical works as a guide in the classification of plants, we may say that most species may be hybridized and varieties cross-fertilized. We cannot say *all* may be, because we would have nothing better than a mere theory on which to base such an assertion, and it is well known that there are many closely allied species that have successfully resisted all efforts to hybridize them. For instance, no one has ever been able to hybridize or produce a hybrid plant between the Alpine Strawberry of both Europe and America and any other of the various species found in different parts of the world. We might also naturally suppose that it would not be difficult to hybridize the different species of the true Cherries (*Cerasus*), but all the species and varieties that produce their flowers in racemes have thus far resisted all efforts to hybridize or cross them with those which bear flowers in umbels. While I would not venture to assert that the various species cannot be hybridized, it is not at all probable that they ever will be. It may be said, however, that while certain species of a genus may have sufficient affinity to admit of hybridizing, others do not. Among some kinds of fruits hybrids have been produced between widely separated and very distinct species, as, for instance, between the Black Cap Raspberry (*Rubus occidentalis*), and the European Raspberry (*R. Idæus*). The late Charles Arnold, of Canada, produced several such hybrids, although they were of no value for cultivation. In view of what has already been accomplished in the way of hybridizing and crossing, and the small number of persons who have experimented in this field, it would be mere presumption on the part of any one to attempt to determine the limits of such operations. It is generally supposed, however, that the hybridizer is restricted in his operations to the members of a genus—that is, he can only produce hybrids between the species of the same genus, and a genus is simply a group of species all possessing similar generic characteristics. But, as I have already shown, the species of a single genus sometimes resist all attempts to force them to intermingle or hybridize, and it is quite probable that they are as far removed from each other in relationship as the members of some distinct genera. It is also probable that in some instances botanists have made mistakes in the grouping of species, as well as in their separation or designation. It is certainly quite possible that two species supposed to belong to a different genera may be forced to interbreed; in fact, hybrids between two supposed-to-be distinct genera are already known, and a hybrid Palm, the *Microphœnix sahuti*, is quite a recent production of this kind. The experimenter need not hesitate to attempt the hybridizing of the species of closely allied genera, although it is quite probable that in a large majority of instances it will be but a waste of time.

CHAPTER IX.

INFLUENCE OF POLLEN.

Whether the influence of the pollen extends beyond the ovule and ovary or not, is a question that has not attracted any considerable attention from either vegetable physiologists or practical cultivators of plants. It is quite evident, however, that there is a reciprocal action extending far beyond the ovary, else unfertilized ovules would not so uniformly show the non-development of the ovary as well as various external organs. In the Rose family, in which we find the Apple, Pear, Quince, etc., the embryo fruit is formed before the flowers expand, and it is the same in the Melon family, and, in fact, in a large majority of plants the seed-vessels and ovules are plainly discernible long before the flowers appear or the sexual organs are developed; but if the ovules are not fertilized, the entire flower and flower-stalk soon wither and drop off. When fertilization occurs, an immediate change in the parts take place; the ovary and its envelope, whether in the form of a pulpy fruit or membranous pod or shell, respond to the demand of the fertilized ovule, enlarging and thickening as it becomes the natural receptacle of the embryo seeds. The flower-stalk supporting these organs also enlarges, as it becomes the vehicle through which nutriment passes from the parent stock to the flower and fruit. The calyx of the flower and other leaf-like organs assimilate the crude sap which reaches them, thereby aiding in the development of the embryo seeds and the seed-vessel. In the absence of fertilization, all the various parts of the flower, embryo fruit and fruit-stalks soon wither away. This is the general result of non-fertilization of the embryo, as every cultivator of plants must have seen. But while it is generally conceded that the presence of the fertilized ovule is necessary to the enlargement of the surrounding and supporting organs, it appears that very few observers have endeavored to trace the influence of the pollen beyond the seed, probably because the practical utility of the operation, in most instances, ceases at this point, although every observing horticulturist must have noticed that the parts most valued in what we call fruits, are at least dependent upon pollination, even if the act does not extend to actual fertilization of the ovule. I use the term pollination to designate an act in plants the equivalent of co-habitation in animals, which does not necessarily extend to or result in fertilization. The necessity of the presence of pollen to produce perfect fruits and seeds is not doubted, but the moot point connected with this matter is, whether or not in cross-fertilization, the pollen has any influence in changing the form, color or properties of the parts enclosing the ovary. My attention was first called to this matter some twenty-five years ago, while experimenting with various pistillate varieties of the Strawberry, the flowers of which, as is well known, must be fertilized with pollen from some perfect flowered variety, in order to secure a crop of berries. When employing varieties bearing large coxcomb-shaped fruit, like that of the Peabody and Triomphe de Gand, for supplying pollen to such pistillates as the Hovey and Burr's New Pine, I noticed that the fruit of the latter two often assumed the shape of the former or pollen-yielding plants. This led me to further experiments in that direction, all of which tended to establish the fact, that the influence of the pollen is often sufficiently potent, not only to change the form, but also the color and flavor of the fruit. I briefly referred to this subject in my "Small Fruit Culturist" published in 1867, and since that time I have had occasion to make many other experiments, for determining the influence of pollen in cross-fertilization and hybridization of different kinds of

INFLUENCE OF POLLEN. 119

plants, and the results, in a majority of instances, have shown that its influence always extends beyond the ovule, but its effect is far from being uniform in producing pronounced changes in the color, form of the fruit, or pericarpic organs. This is not at all strange, inasmuch as the plants bearing the fertilized flowers also supply the entire nutriment necessary for their support and development, hence, would naturally have a much greater influence on the growth of the seed-vessel than any likely to be conveyed in a few grains of pollen. Still, the influence of the pollen on the female organs is readily discernible, and through these it must necessarily affect, to a greater or less extent, all parts of the plant that respond to the demand for nutrients to the seed and surrounding organs. We can readily see the effect of fertilization, and often mere pollination, in plants by the rapid swelling and growth of the pericarpic organs—the fruit stalks and their various appendages—and even trace it down to the very roots, of such plants as the Strawberry and other herbaceous kinds.

Pollen is not a simple substance, but a compound, and while its principal office is to convey to the female cells fertilizing materials, it may also carry elements of health or disease, as well as those that are or may become hereditary characteristics of its race. In the animal kingdom we have an exact parallel case, for, leaving out all mental impressions, the male parent contributes no more towards the production of the offspring in proportion to size, and we doubt if as much, as does the male in the vegetable kingdom. Still, no breeder of choice stock would willingly permit the contamination or adulteration of the blood of the female by a scrub animal, or even one out of the direct line of the pure breed. It is an old saying that "there is a black sheep in almost every flock," but there are probably few persons who know the full significance of this phrase when applied to sheep, and especially to the long known American Merinos. No breeder of these sheep in the past fifty or sixty years has allowed a black male in his flock, and yet almost every season an ewe among the pure bloods, as they are called, will drop a black lamb, a direct reversion to some early ancestor of this color. No one can tell how long, or through how many generations, this early contamination will continue to show itself. Similar cases of reversion, the result of contamination of blood, as it may be termed, are known to frequently occur among all of our domesticated and improved breed of animals, and they are as likely to originate with and become hereditary in the male line as in the female. As there is an affinity between animals to admit of breeding, so there must be the same among plants, and the mobile pollen promotes, if it does not cause, excitability in the female organs of the flower, and thus produce a responsive action from those below and beyond, as already stated.

While it is not claimed that the influence of the pollen will, in all cases, show itself in marked changes in the form or size of the pericarpic organs, still that it often does this can scarcely be doubted by any one who has ever made careful experiments for the purpose of ascertaining the truth in this matter. In cross-fertilizing varieties of Indian Corn of different colors, the influence of the pollen can be readily traced, not only in producing kernels on the same ear of different sizes, shape and color, but the cob underneath the black, yellow or red kernels will usually be tinged with a corresponding color. In all plants having a distinct pistil for each seed, as in Indian Corn, the Sorghums and Millets, or in fruits like the Strawberry, Raspberry, etc., each and every ovule must be fertilized in order to produce a perfect ear or fruit. If none of the ovules are fertilized, the ear, fruit, or pericarpic organs, and even supporting fruit-stalk and stem, wither away.

Since the attention of horticulturists has been called to this subject, several who combine science with practice have repeated my earlier experiments with the Strawberry, and in a majority of instances with like results. Prof. W. R. Lazenby, of the Ohio Experiment Station, made several very carefully-conducted experiments with different varieties of the Strawberry in 1884, the results of which were fully reported in the Bulletin of the Station for that year. In referring to these experiments at the meeting of the American Pomological Society in 1885, Prof. Lazenby stated that he employed a pistillate variety of the Strawberry, known as the Crescent. "Boxes open above and below, and covered with whitewashed glass, were placed over the plants to prevent accidental fertilization of the flowers by insects or otherwise." The results in brief were: The Charles Downing pollen communicated its characteristic shape, texture, and other qualities, and the same with Sharpless and Vick, so that any one familiar with the berries could tell by looking at the crop from what source the pollen came. The following season, or in 1885, Prof. Lazenby repeated the experiment, but with far less satisfactory results, showing, as I have said, that the influence of the pollen is not always sufficiently powerful to change size, color, etc., or it may be said that the influence of the female plant is strong enough to overbalance and partly suppress that of the male.

Prof. Julius Sachs, in his recent great work, "Text Book of Botany," says, p. 495: "The increase in size of the ovary, which is frequently enormous (in *Curcubitæ*, *Cocos*, etc., several thousand times in volume), shows, in a striking manner, the results of fertilization * * * to the rest of the plant. Frequently similar changes result also in other parts. Thus, it is the receptacle that constitutes the fleshy swelling which is called the Strawberry, on the surface of which are seated the small true fruits." Also on page 594: "But sometimes the long series of deep-seated changes induced by fertilization extend also to *parts which do not belong to the ovary* and even to some *which have never belonged to the flower.*" Among the plants so affected he names the Fig, Strawberry and Mulberry.

In some kinds of plants even the ovary itself appears to be the result of pollination. Dr. Hildebrand has shown that in some of the Orchids, that it is only during the growth of the pollen tubes through the tissues of the stigma and style that the ovules become so far developed that fertilization can at length be affected. In other words, in the absence of pollen, no ovary or organs for containing ovules or seed are produced. But as the space at my disposal will not admit of reference to the various authors who have touched upon this subject, I must omit them, for my principal object in referring to the influence of pollen upon other organs besides the seeds is to open the way for seeking an explanation of the cause of certain phenomena often occurring among cultivated plants—for instance, the occasional change of color and general appearance of specimens of fruits, when growing in close proximity to other closely allied but distinct species or varieties. There are many instances on record, and others are being noted every season, both in Europe and this country, of fruits, and especially Apples on one tree, assuming the color of those of an adjoining tree. Sometimes only one or two specimens on interlacing branches are thus affected, while in other instances nearly all, or a certain branch, will show the effect of the cross-fertilization. Several such instances were observed and reported by Dr. Hildebrand, of England, to the *Gardeners' Chronicle*, some twenty years ago, and in each case the change in the color of the fruit showed very clearly that it was due to the influence of the pollen of flowers on branches of other varieties growing in close proximity. White Apples have been colored when growing near red varieties, and those having smooth skins have been covered with russet on branches that have been interlaced with those of russet trees. That this cross-fertilization does not occur every season, is due to the fact that self-fertilization is the most natural and frequent in the flowers of the Apple and kindred fruits, and further, when crossing does occur, it is not to be supposed that the effect will always be sufficient to change the color or other characteristics of the fruit.

In the many experiments that I have made for the purpose of determining the extent to which the influence of the pollen could be distinctly traced, I will only refer to one which has recently been completed on plants, so common that the merest tyro in gardening must not only be acquainted with them, but can readily repeat my experiments if they care to do so. I refer to the Shrubby Abutilons, which come to us from South America, where some of the species grow to a height of thirty feet. Nearly all the species and varieties in cultivation bloom freely either in the open ground or under glass, but produce seed very sparingly, and some none at all, unless artificially cross-fertilized. Another point in their favor for testing the influence of pollen is, that the seed-pods of the different species and varieties are quite various in size and form, and any disturbing or foreign influence can the more readily be traced to the pericarpic organs. Among the twenty or more species and varieties in cultivation, I selected *Santana* and *Boule de Neige* as two extremes in the way of varieties likely to be the best for my experiments. These may be only varieties of the same species, but from the form and color of these flowers I am inclined to think that *Santana* is from *A. venosum*, and *Boule de Neige* from *A. pulchellum*. My plant of the *Santana* is seven or eight years old, and is planted in the ground in one corner of my greenhouse, and has been cut back several times to keep it within bounds. It is a mass of flowers all the year round, but has never shown a a sign of producing seeds, except as the flowers are artificially fertilized by pollen from some other variety. Its own pollen is impotent for fertilizing its own stigma, as I have proved by repeated experiments, but it is perfectly potent on other varieties. The form of the seed-pods of the *Santana* is broad top-shaped, the ten to twelve loculicidal cells are each tipped with a prominent winged terminal point, as shown in figure 48, while the seed-pods of *Boule de Neige* are slightly smaller, and the cells, instead of widening at the top, are contracted, or more or less rounded inward, as shown in figure 48 at *A*. Now, to fertilize the stigmas of the *Boule de Neige* with pollen from the *Santana* is but a simple operation, requiring only a few minutes for its performance. Twenty flowers in one experiment were thus fertilized, and the result was that five of the seed-pods assumed the form of those of the *Santana* or male parent, as shown in figure 48, *B*, the other fifteen retaining the normal form. The plants, however, since raised from the seed show plainly the effect of cross-fertilization. But in reversing this operation, and in fertilizing more than fifty flowers of the *Santana* with pollen from those of the *Boule de Neige*, in no instance has there been any very marked change in the normal form of the seed-pods, but I have seldom failed to effect cross-fertilization or obtain fertile seed. The *Santana* being much the strongest and most vigorous plant of the two operated upon, it is not at all strange that what I presume is the normal form of the seed-vessels should be retained under the circumstances. As a rule the female plant, if equally as vigorous and healthy as the male, will exert the greatest influence upon the offspring, because there are no

Fig. 48.—BOULE DE NEIGE ABUTILON, CROSSED WITH SANTANA.

mental faculties to assist in impressing and perpetuating the male or other sexual characteristics, as in the animal kingdom.

As I have shown in the case of the Abutilons, sexual affinity presents wide variations, which can only be definitely determined by actual experiments upon the plants themselves. No theory will explain why the pollen of a plant cannot fertilize its own ovules, while it has an affinity for, and is potent upon, those of another closely allied plant. If the pollen from several different varieties or species is applied simultaneously to the same stigma, it is quite evident that only one kind will be potent, and that one from the plant possessing the greatest sexual affinity with the plant pollinated.

It is generally supposed that in all the higher orders of plants the ovule must be fertilized in order to secure perfect and fertile seed, but there are some few instances on record where seeds are supposed to have been produced in the absence of fertilization, or, as it has been termed, by *Parthenogenesis*. Prof. Asa Gray thinks it does sometimes occur among Diœcious plants. A century ago (1786) Lazaro Spallanzani, published his observations on the fecundation of plants, and claimed to have found pistillate blossoms of the Hemp producing fertile seed in the absence of pollen. A half century later Drs. Charles Naudin and Joseph Decaisne are said to have confirmed the fact by raising seedlings from Euphorbiaceous plants, also from the common Bryony, which were kept from all access of pollen. While I am not disposed to question the statements of such high botanical authorities, or to attempt to offset their experiments with those of my own, still I think there is room for a doubt in this matter, and especially when we take into consideration the present undetermined boundaries of species.

The question arises, may not the pollen of some other

and closely allied plant have reached the stigma, causing a sufficient responsive action to insure fertile seed? Pollination does not necessarily extend to fertilization, and the presence of pollen on the stigma may result in pollen tubes, which, upon reaching the ovule, fail to fertilize it, owing to the want of sufficient natural affinity between the two. In plants like the Strawberry, the pistillate varieties are merely abnormal productions, the result of some chance suppression of the stamens, and in such instances we may always expect an occasional development of the suppressed organs. It has been said that "necessity knows no law," and there are good reasons for expecting that nature will make the effort, and she is often successful in supplying an absent organ, as well as employing various means to accomplish the same results.

My own experience leads me to doubt the production of fertile seeds where there is a total absence of pollen, either in diœcious or other highly-developed plants, but it may not be always necessary that the pollen should be supplied by plants of the same or very closely allied species. For instance, for many years I had growing in my grounds a pistillate specimen of the Box Elder or Ash-leaved Maple (*Negundo aceroides*). There being no staminate or male tree in my own grounds or in the neighborhood, at least not within a radius of six or eight miles, consequently a fertile seed on the said tree could not be found. Every year for a decade this tree was loaded with its unfertile seed, which were scattered broadcast under the branches of some large pines—a favorable position for germination if there had been a perfect seed produced among the vast number. There were many other species and varieties of the Maple in my grounds, but none growing very near or that bloomed at the same time as the Negundo. In the spring of 1879 I planted a large Red Maple (*Acer rubrum*), about sixty feet from the Negundo, and the transplanting delayed its blooming, the flowers appearing with those of the former, although the Red Maple usually blooms several days earlier than the Negundo. The next season I found several seedling Negundo Maples coming up under a Pine tree near by. These were transplanted, and last spring, 1886, one of them bore seed, a pistillate like the parent. While I cannot say positively that the flowers of the Negundo Maple were pollinated or fertilized by those of the Red Maple, still the facts in the case point strongly that way. The seedlings produced in this anomalous manner do not appear to be hybrids, but this point cannot be definitely determined until plants are raised from the present race of seedlings, and even then the influence of the pollen-bearing parent may remain hidden through several generations and afterwards appear when and where least expected. It is not at all improbable that the presence of non-related pollen may excite development in the pericarpic organs, and if this be true we can the more readily account for the production of fruits with abortive seeds—the stigma having been pollinated, but the pollen tubes having failed to fertilize the ovules.

The excitability of plants due to the presence and influences of pollen has long been observed by the cultivators of the Hop plant, the quantity of lupuline deposited in the strobiles or female catkins being far more abundant when there are staminate or male plants present than when they are absent. It is also well known that among our larger fruits we have seasons of scarcity and of abundance, and these variations cannot be attributed to changes in the climate or to age, maturity or non-maturity of the trees, for planting is going on continually, so all sizes, ages and varieties are represented in our orchards, still all seem to readily acquire the habit of bearing full crops every alternate year or at longer intervals. That this unity of action among fruit trees, as well as forest trees, may in part be due to exhaustion in bearing a full crop, and the necessity for a season or two of rest and recuperation, is probably true, as some have claimed; but that it is largely due to general periodic sexual excitement can scarcely be questioned, for similar phenomena are of frequent occurrence in the Animal as well as in the Vegetable Kingdom.

PURPOSES OF CROSS-FERTILIZING.—The results sought in cross-fertilizing of varieties, or the hybridizing of species, are various, but the principal one is to produce something different from either parent. Sometimes we may aim to increase the size, or change the color, texture, flavor or other characteristic of a fruit, or the size, form and color of the flower or habit of a plant. Adaptation of the various species and varieties of cultivated plants to specific conditions is another, and often a very important, object sought in producing cross-bred plants. There are many species, and occasionally varieties, which have been so closely and continuously inbred in their native habitats or elsewhere that they have acquired a fixedness of character that removals to other localities, and subjection to widely different conditions fail to effect any material change in their offspring; but by crossing, and the introduction of new sexual or other elements, the foundation of generations, as it may be termed, is broken up, and wider deviations from normal types soon follow. It may sometimes be necessary to introduce an undesirable element, in order to force a plant to break away from its typical form, but when we have succeeded in this, it will not be difficult to breed out the undesirable characteristics or properties. Then, again, we cannot know in advance what will be the result of crossing any two plants of the same genus or species, for even the mingling of two inferior elements may result in the production of one superior to either of the originals; still, we would not advise using inferior materials in preference to superior, except when it is absolutely necessary to effect a desired variation, as may sometimes occur when a person has but a limited number of species or varieties of a genus with which to experiment.

A wilding may possess some very desirable properties, such as vigor, hardiness and exemption from disease, while its domesticated congener lacks one or all of these properties ; so, by combining the best elements of the two, a new and superior progeny may be produced.

As yet no very marked improvements among cultivated fruits have been made by hybridizing, although in a few instances, as with the Grape, it may have aided in causing the species to break away from the wild type, and through the wide variation resulting from artificial fertilization some valuable varieties may have been secured. Whether the hybrids between the indigenous species of the Grape of North America and the *Vitis vinifera* of Europe are as well adapted to our climate as the pure native varieties is at least questionable; still, it may be that the introduction of a foreign element will yet prove to have been a judicious movement, and in the right direction, for yielding the best possible results.

The object in all cases should be to introduce valuable properties, and in such a combination that they can be made available. We may, among fruits, secure size, color, texture—in fact, all the good qualities known to belong to or exist in a certain species, and still these will be of little value unless the plant itself is adapted to the soil and climate where it is to be cultivated. In fact, adaptation is all that is sought or can be credited to what is termed acclimatizing of plants and animals, for it is scarcely to be supposed that the constitutional characters of the individual plant or animal can be greatly or permanently affected by a removal from one climate or condition to another. One variety of plant may be more hardy, and safely endure many degrees lower temperature, than another of the same species, but no amount of nursing or moving about will ever change a tender plant or animal into a hardy one. But by introducing new elements, as in cross-fertilization, we multiply the causes for wide variation through the different hereditary characteristics of both parents. Then, by careful selection and propagation of such cross-bred varieties as are worth preservation, we are often able to secure those adapted to widely different conditions, as seen among all of our long-cultivated and widely-disseminated plants. Why the seeds from a plant should yield both tender and hardy varieties can only be accounted for upon the hypotheosis that each possesses transmitted hereditary characteristics, but what the nature of the laws are that control this transmission we know little or nothing.

Plants that are indigenous, or have become naturalized in cold climates and in elevated regions, are constantly subjected to the loss of their leading shoots and branches through the action of frosts and winds, and this being repeated for centuries in succession, the plants at last become permanently dwarfed, and this character becomes fixed and hereditary, as seen in many of our cultivated plants obtained from very cold or alpine regions. The dwarf habit remains fixed in the individual, even when cultivated in more favorable regions of a country, but so soon as we commence to raise seedlings from these pigmies, we find that there is a tendency in a certain number to grow taller than the parent plant, or to return to what we may presume was the original form of the species.

Form and habit of plants are greatly modified by surrounding conditions, and while severe winds and low temperature may permanently dwarf plants in certain countries or regions, high temperature and poor soils may produce similar changes elsewhere. The Chinquapin Chestnut (*Castanea pumila*), as found growing wild over quite extensive regions in some of our Southern

States, is a familiar instance of a tree being reduced to a mere shrub through the influence of a poor soil and other uncongenial surroundings. Under favorable conditions this species grows to a tree thirty or forty feet high, but where they are unfavorable, it is but a mere shrub from two to six feet high, even when the plants have reached what may be considered maturity.

Instances of permanent changes having been effected through various external causes are no doubt familiar to every botanist and gardener, but the practical plant grower should be on his guard, lest the sudden appearance of some hereditary character, or direct reversion in seedlings to some ancient type, mislead him into thinking that the variation observed is the result of his own skill in cross-fertilization.

All plants have a tendency to vary in different directions, the cultivated more than the uncultivated, while every new internal element introduced, intensifies this proclivity to depart from the normal type, and whenever a departure has been made, it is likely to become hereditary, consequently it is often difficult, if not wholly impossible, to determine to what disturbing cause we are indebted for certain results.

BUD VARIATION.—This is a prolific source of varieties among plants, although by far the greater number are raised from seed. When a bud on a tree or other plant from some cause unknown, produces a shoot or branch differing from others on the same stock, it is attributed to what is called "bud variation," and the branch or shoot so produced is termed a "sport," to distinguish varieties originating in this way, from those raised from seed, or the direct product of sexual reproduction. These sports, if removed from the parent plant and propagated by division, will often remain permanent, but sometimes they quickly revert to the original or parent form. In propagating varieties originating from bud variations, it is usually necessary to exercise considerable care, in order to preserve their distinct characteristics, and this is especially true in those with variegated leaves, as seen in the variegated leaved Elder, Dogwood, Ginkgo, Maples, etc., for all of these possess a strong tendency to revert to the plain or one-colored leaf of the parent. Through bud variation we have obtained many of the most highly prized ornamental trees and shrubs, both evergreen and deciduous, and new ones are constantly being added to the list. There is certainly a cause for the variation of buds. In some plants it may be an attempt to revert to some earlier form, and in others the result of some element introduced through the sexual organs, either in the present or some previous generation; but whatever the cause, it is certainly an inherent property of the plant which is not confined to one part or single bud, for it frequently occurs that several buds on the same plant, but on different branches on distant parts of the stem, produce shoots possessing the same characteristics. A noted instance of this kind occurred a few years ago on the Remilly Ash, growing near Metz in France. This tree is over sixty feet high, with stem about six feet in circumference. The branches are pendulous, and the variety is known as a Weeping Ash; but three buds, one on the main stem near the top and two on separate branches lower down, produced shoots which assumed an erect habit, and have continued in this until they have become large branches. Similar freaks or bud variations have been frequently noticed on other weeping trees both in this country and Europe. When a plant has shown a tendency to produce these bud variations or sports, a repetition may be looked for, even if the first one is promptly removed.

About ten years ago I noticed a single branch on one of my specimen plants of Golden Retinispora (*R. obtusa, var. aurea plumosa*), which had assumed an entirely distinct form of growth from all others on the same plant, or in fact on any of the various species or varieties of the Retinisporas in my grounds. This branch was layered, and the next season cut off and planted out by itself. It is now more than three times the size of the parent plant, and so distinct from it in leaf and habit of growth that it might readily be taken for a distinct species instead of a sport. Prof. George Thurber, the eminent botanist, described this sport in 1881, and named it *Retinispora obtusa Fulleri*. Three years ago another bud on the same plant, but on the opposite side of the stem, produced a branch identical in every respect with the first sport, showing that this plant has either inherited or developed a faculty of producing bud varieties, and of one form only.

Sometimes these bud varieties show a marked difference in the color of the foliage or habits of growth, while in others only the flowers appear to differ, as seen in many well-known varieties of the Rose which have originated in this way. The first Moss Rose is supposed to have originated from a bud on the old Centifolia; the Striped Moss is a sport of the old Red Moss, and the White Baron Prevost from the old Pink Hybrid Perpetual of the same names. The American Banner, Triomphe d'Amens, Painted Orleans, and many other well-known varieties of the Rose, originated from what are termed bud variations. Among fruits bud variations are constantly occurring, but the larger number are probably overlooked and consequently lost. The Red Magnum Bonum Plum is said to have originated from a bud of the Yellow Magnum Bonum. Many instances are recorded of Peach trees producing Nectarines on one or more branches, and these sports have frequently been preserved and extensively propagated. The seed of Nectarines originating in this way usually produce Nectarine trees, not reverting to the Peach.

The earliest instance on record of a Peach tree producing Nectarines is mentioned by Peter Collinson, in 1741, but they have been so frequent since that they have ceased to be considered rare or strange. Similar bud variations may be looked for among all of our cultivated fruits, and the horticulturist should be constantly on the alert for such freaks and endeavor to perpetuate those likely to prove valuable. What is true in fruits is also true among ornamental trees, shrubs and herbaceous plants, and those who are seeking new and valuable varieties may often find them where least expected.

CHAPTER X.

GENERAL PRINCIPLES AND METHODS.

The different modes of propagating plants may be classed as follows :
1.—Propagation by means of seeds and spores.
2.—By cuttings of the stems, twigs and leaves.
3.—By suckers and divisions.
4.—By root cuttings.
5.—By budding and grafting.

All the known species and varieties of plants may be multiplied by one or another of these methods, and some kinds can be readily propagated by each and every one of them. But there are certain principles, which serve as a guide to the propagator in the different modes of operation, that it may be well to consider before proceeding to the more practical part of the subject. Although, when the great diversity of characters as well as the vitality of plants is considered, it cannot be expected that any general rule can be given that will be applicable to every case or be altogether faultless, yet for the purpose of dispelling that mystery with which the novice often supposes the various modes of propagation to be surrounded, I shall give a brief synopsis of the general principles connected therewith, together with a description of the more usual methods practised by our most skilful propagators of plants. It may also be proper to suggest that, however well informed a person may be in regard to the structure and habits of plants, and however extended his experience and perfect the conveniences for propagation, he must still possess skill and patience, and exercise great care and watchfulness in every operation, in order to become a successful cultivator and propagator of plants in general. A person may know just how an operation should be performed and still lack the skill required for its execution.

PROPAGATION BY SEEDS.—The perpetuation of the greater portion of the known species of plants is directly by their seeds, which, in their wild state, they perfect with great uniformity, but when cultivated the vital forces are often disturbed, and a portion is directed into other than natural channels. The effect of this we see in the double flower and the increased size of many of our fruits. The seeds of so-called "improved plants" often become abortive and cannot be depended upon as a means of reproduction, not only because of the want of vitality which naturally belongs to them, but in the course of long cultivation there has been such an intermingling of species, as well as varieties, that scarcely any variety of cultivated fruit will reproduce an exact counterpart of itself from seed. Therefore we have to resort to other modes of propagation to perpetuate any particular variety.

Under what particular conditions seeds germinate most readily it is certainly difficult to determine, because of their great diversity of character and functions. Heat and moisture are always necessary in order that certain changes may take place in the seed, but the degree of heat and amount of moisture required are exceedingly variable. As stated elsewhere, some seeds will germinate at a temperature slightly above that of freezing water, while others require nearly or quite a hundred degrees Fahrenheit. Certain kinds of seeds will absorb more than their own weight of water before sprouting or bursting their hard, bony covering, while others appear to retain, while ripening, nearly all the moisture required in the early stages of growth. Seeds will withstand a higher and a lower temperature in a dry atmosphere than in a moist one, hence cool and dry, as well as hot and dry, are non-inciting and preservative conditions for most kinds of seeds. But seeds in general are not so frail as to be readily destroyed by slight variation in temperature, or in the hygrometric conditions of the atmosphere, and their germination may usually be hastened or retarded without serious injury, although it is always safe to place seeds under conditions approximating those under which we find them in their native habitats, but in order to do this we must know something of their history. The spores of Cryptogamous plants indigenous to a cold climate might quickly perish if placed under the conditions most favorable for the growth of a Cryptogam from the tropics, and the same difference may often be noticed in the seeds of higher orders of plants. With a knowledge of the peculiarities of the climate of the habitat of plants, the propagator can usually make a very close guess in regard to the treatment the seeds require to insure germination. It is well, however, to bear in mind that latitude and longitude do not furnish a very trustworthy guide to climate, for there are snow-capped mountains and very cool or cold regions of country even in the tropics, or within the belt of country so designated by degrees of latitude. Ocean currents also have their influence, and the climate of widely separated regions, on the same parallel of latitude, may differ greatly, as seen in the difference between that of New York and that of the City of Madrid in Spain, or in the climate of England, and of Labrador. Then, again, the hygrometric conditions may have a very marked influence on vegetation, as seen in the cool, dry, elevated regions of our own and other countries. Plants, native of cold, dry climates, often fail in moist, warm ones, while others are wonderfully improved by the change, and all these varying conditions and results must be duly considered in their propagation by seeds or otherwise.

When supplied with the requisite amount of heat and moisture, all seeds grow more readily when near the surface of the soil than when buried deeply, therefore we should endeavor to so place them in or on the soil that the air and heat can reach them, but at the same time exclude the light, for darkness is favorable to germination. But air, or rather oxygen, is necessary, consequently deep planting and exclusion of air retards or wholly prevents growth. Whether seeds should be covered or not with soil to assist or insure germination depends much upon circumstances, as well as upon their size, structure and power of throwing up their stems through the material used for covering them. Some kinds of seeds will germinate far more readily if placed on the surface of the soil and then kept moist and in shade or total darkness. It may be said that we only bury seeds as a matter of convenience, and not because it is actually necessary to insure germination. By covering seeds with soil or other similar material, we are enabled to secure more equable conditions as regards temperature and moisture, as well as the exclusion of light, than if they be left uncovered, and thus we secure better and more uniform results with less attention ; still, roots will generally penetrate to the required depth and position even when the seed has not been buried, provided they fall upon some yielding material or one readily penetrated by the young rootlets. Seeds of our common forest trees, when they fall upon the thick mat of old leaves, usually fail to grow because the young roots cannot penetrate through the tough, fibrous material underneath them; but let these seeds be scattered along the roadsides, or in the open fields, or wherever they can come in direct contact with moist soil, and they soon show that they have found congenial conditions for growth.

No general rule can be given in regard to the depth for covering seeds, not even one that would apply to all the members of the same family of plants, because they frequently differ in their habit of growth, as, for instance, in Peas and Beans, the cotyledons of the former remaining under ground and those of the latter being lifted above the surface in the elongation of the stem; consequently we can safely cover a small variety of the Pea much deeper than the largest variety of the Bean, although both are closely allied dicotyledonous plants. Similar variations also occur among the monocotyledonous seeds, the plumule or stem usually rising in the form of a cylindrical column, whether they are of the size of the giant Cocoanut, or of Rye, Wheat, or the still smaller grasses. Depth, however, must be varied somewhat, according to the nature of the soil in which seeds are sown, for a stem that would readily push up through an inch or two of vegetable mold or of sand, might be unable to pierce the same thickness of compact loam, or tough, hard clay.

When seeds have once begun to grow, they cannot be again reduced to a dormant state without causing their destruction; this should always be borne in mind, for from this cause alone more seeds are annually destroyed than from any other. As they are usually hidden in the soil, we are very likely to neglect giving them an ample supply of moisture at their time of greatest need.

The soil in which seeds are sown should be made fine

and readily permeable, not only to admit air and heat and to retain moisture, but so that the radicle or young root may penetrate the earth without hindrance, and also permit the stem to grow upright unimpeded. The soil should in most cases be deep, and of such a nature that the young plant will receive a constant and regular supply of moisture. Nature serves as a guide to us in many of the operations connected with the art of propagation; yet it should be remembered that the sowing of seeds and transplanting are artificial operations instead of natural ones, and we follow nature only when it serves our purpose, or as we are compelled to do so by natural laws. Nature is so plastic that she allows us to mold her gifts into forms that meet our wants and tastes, confining us only within certain limits that are difficult to determine.

Nature perfects as well as destroys, and thus that equilibrium is preserved which is observable throughout the vegetable kingdom. If we scatter seed in every instance exactly as is done by nature, we should not make more than one in ten thousand grow.

While it is quite obvious that seeds require moisture to insure their germination, it is only the aquatic kinds that will bear an unlimited amount; for, while the seeds of ordinary field and garden plants may sprout when submerged, they require air to insure a continuous growth. To keep a seed-bed constantly soaked or saturated with water is almost as injurious as to allow the seed to suffer for the lack of a sufficient amount of moisture. The propagator must always exercise his own good judgment in such matters, varying his treatment according to the size and nature of the seed as well as the kind and condition of the soil.

Seeds forced in a high temperature, or sprouted in a low one, usually produce feeble plants. If the temperature of the soil is between fifty and sixty degrees, Fah., it will be favorable for the germination of seeds of plants indigenous to cool climates; but a temperature of sixty to seventy degrees will be none to high for ordinary greenhouse plants; while seeds from the tropics may need ten or twenty degrees higher temperature.

While it is generally conceded that new seeds are preferable to old ones, still the idea of newness should not mislead the propagator and cause him to employ immature seed; for, in many instances, what would be considered a fully ripe seed has not arrived at its best condition for sowing or growth. Seeds containing a large amount of natural moisture, or sap, will often fail to grow if placed in the ground when new and fresh; but if left to dry a few weeks or months and then sown, they will germinate freely. I am unable to give any scientific reason for these variations, but have learned from long experience that while freshly gathered seeds of some kinds of plants will fail to grow, after having been dried for some time, they germinate very readily. It would appear that the loss of natural moisture and shrinking increases the power of absorbing external moisture, and accelerating the chemical changes that take place during germination. But if the drying is allowed to proceed too far, the vital energies are diminished or wholly destroyed. A period of rest appears to be necessary to the seeds of a large majority of plants, while, on the contrary, there are kinds which must be placed in a position for growth before losing much of their natural moisture, else they fail to germinate. For instance, the seeds of our indigenous White Maple (*Acer dasycarpum*), and those of the Red Maple (*A. rubrum*), will not withstand drying, and must either grow within a few days after falling from the tree or perish. But of the seeds of the Elm (*Ulmus*), of various species, ripening at the same time as the Maples named, and also similar in structure, some will sprout immediately, while others remain dormant until the following year, although all may be treated exactly alike. The same variations occur among many different kinds of seeds, some germinating readily when freshly gathered, others requiring a season of rest, or time to become fully mature. It is well known that there are periods of great activity and rest in the vegetation of all countries and climates, and these have their influence upon the seeds, their vital energies becoming excited with the return of the season of growth, and at such times they will germinate far more readily than at any other, although artificial surrounding conditions may appear to be the same during the period of growth and when they are at rest.

To hasten the germination of seeds, gardeners have recourse to various expedients. Steeping in different solutions is often practised with old seeds, in order to soften the outer covering and admit moisture to the interior and germ. Seeds with hard, horn-like integuments, like those of the Three-thorned Acacia (*Gleditschia triacanthos*), and Kentucky Coffee tree (*Gymnocladus Canadensis*), may usually be forced to germinate by steeping a few days in warm water, or hot water may be poured over them and allowed to cool to a temperature of about 100 degrees, and kept at this point until the seeds show signs of growth. Seeds incrusted with resin, as is usual with the Junipers, or wax, as in the Bayberry (*Myrica*), are benefited by steeping in a solution of potash, or they may be mixed with moist wood-ashes, and kept in this condition until the incrustation is removed and the bony nuts are somewhat softened. Potash solutions are to be recommended as a steep for all similar seeds, especially if they have been allowed to become dry and hard, or when it is desired to force their germination in advance of their natural season of growth. Lime is also used for similar purposes, and a few years ago it was highly recommended by several European horticulturists for hastening the germination of old Spruce seeds and of other conifers. Humbolt is said to have employed a dilute solution of chlorine with great success in sprouting seeds; and another German, a Mr. Otto of Berlin, employed oxalic acid to make old seed germinate. Dr. Lindley, in his "Theory of Horticulture," describes Mr. Otto's method of using oxalic acid, and says that "the seed were put in bottles filled with the acid and left in it until they germinated, which generally takes place in from twenty to forty-eight hours." The seeds are then removed and sown in the usual way. Very small seeds may be sown and diluted acid applied two or three times a day until they germinate.

Mr. Otto claimed that by this means, seeds that were from twenty to forty years old grew, while the same sort, sown in the usual manner, did not grow it all. But Dr. Lindley, in referring to the advantages claimed for the acid process, says: "Theoretically it would seem that the effects described ought to be produced, but general experience does not confirm them; and it may be conceived that the rapid abstraction of carbon by the presence of an unnaturally large quantity of oxygen may produce effects as injurious to the health of the seed as its too slow destruction in consequence of the languor of the vital principle."

The oxalic acid solution for accelerating the germination of seeds, although highly extolled at the time of its discovery, has, like many other similar discoveries, gone out of use and is almost forgotten. At the present day our horticulturists depend mainly upon heat and moisture for revivifying the dormant energies of their old as well as new seeds, although, as I have said, the alkaline solutions are very useful in softening and removing natural incrustations of some kinds, and the hard, horn-like covering of others. Frost, or freezing, is also useful for this purpose, and always available in cold climates; but heat answers the same purpose for opening the pores and admitting moisture to the seed proper ; and while in cold climates, nuts and other hard shelled seeds are placed where they will freeze in winter, the same kinds sprout just as freely in hot climates, provided they are kept moist and warm during the same season.

CHAPTER XI.

PROPAGATION BY CUTTINGS.

OF MATURE GROWTH.—There are many kinds of trees and shrubs that are readily propagated by cuttings of the mature or ripened wood. Sometimes wood two or more years old is used for this purpose, but with most kinds that of one season's growth produces roots more readily. The cuttings are usually taken from the parent stock in the fall of the year, as soon as the leaves of deciduous plants will part from the stem without injury to the buds adjacent.

Autumn is also a proper time to make cuttings of many kinds of evergreen plants, particularly those of hardy trees and shrubs indigenous to temperate climates. A branch, when it ceases to grow in summer or autumn, contains a large amount of matter which has not assumed any special form or structure, and it is therefore in a proper condition either to produce roots or branches. With some kinds of plants it can be made to produce the former very readily; with others, it is quite difficult, simply because we have not discovered the proper condition necessary for their development; and it is just here that we come upon the great secret in the propagation of plants—*i. e.*, under what conditions should cuttings of a particular plant be placed to insure growth? Cuttings of the Willow, Currant, and many other woody plants, grow very freely, even if taken from the parent stock at almost any time of year; while it would be difficult to make a branch from a Hickory tree produce roots under the most favorable conditions; yet it may not be among the impossibilities to propagate the Hickories from cuttings. The horticulturist, however, does not usually seek the most difficult methods of multiplying plants, but the easiest, and there may be many ways of producing the same result.

As it requires more or less time for a cutting to produce roots, it is better to allow an abundance than too little; consequently, we usually make the cuttings of the mature and dormant wood in the autumn, because by doing so we secure several months in which to produce the change; or, in other words, for roots to form. Roots are produced readily at a lower temperature than leaves—also in the dark—and these conditions are easily secured, even in cold climates; for if the earth is frozen on the surface, it may still be warm enough below to afford sufficient warmth to insure the formation of roots on the cuttings of woody plants native of a similar climate. We avail ourselves of the knowledge of this fact and make the cuttings of hardy plants in the fall, and either plant them immediately where they are to grow, giving further protection if necessary, or bury them in a cellar or the open ground; in fact, almost anywhere that we can secure a temperature but slightly above the freezing point, but not so warm as to force the leaf-buds into growth.

In such situations the process of forming roots will go on, and some kinds will usually become so well supplied with roots by the time the regular growing season commences in the spring, that a vigorous early growth of stem will be produced. These conditions are produced naturally in the open ground, for the temperature of the soil in spring is generally warmer than the atmosphere, and the lower end of a cutting, from which point it is always desirable to have the roots produced, receives more heat than that portion which is exposed to the air. In latitudes where the ground freezes to a considerable depth, every one who has ever taken the trouble to examine the soil at the time of its thawing out in spring, must have noticed that it thaws from below upward, far more rapidly than from the surface, downward. Heat descends slowly, but cold rapidly, and just as soon as the weather becomes so warm that the surface does not freeze, the heat from below will rise to the surface. The hot-bed used by gardeners in forcing vegetables in cold climates is made on the same principle, the object being to secure warmth for the roots, while the leaves and upper portions of the plant are kept cool. Thus plants are forced with what is termed "bottom heat."

In warm climates it is just as important to give the cutting plenty of time to form roots, or the advance process called a *callus*, as in cold ones; for if roots are not formed when the leaves expand, the cutting is very likely to die. The callus which always precedes the formation of roots on all kinds of cuttings, whether from ripe or green wood, leaves or roots, is composed of cellular matter formed principally from the assimilated sap of the plant, and is of a similar nature to the nutrients stored up in seeds for nourishing the embryo and young plantlet; and while the callus does not itself become a root, it is the immediate source from which the young rootlet on a cutting obtains nutriment. By a close examination of the callus on a cutting at the time the young rootlets are pushing out, it can readily be seen that the roots are distinct formations, and not in any manner the result of an unfolding or prolongation of the irregular masses of exuded matter which is termed the callus.

Cuttings that are removed from a cellar or other place,

where they have been stored during the winter months, to the open ground, need to be handled with great care, especially if well callused and young roots have commenced to push out; for any considerable exposure to the light, to drying winds, or to rough handling, will check the root-forming process, if it does not entirely prevent further progress toward growth. In taking out cuttings of this kind for planting, they should be laid carefully in rather shallow boxes, kept shaded, and occasionally sprinkled if necessary to keep them moist while being set out. If kept moist or wet the soil will adhere to them closely, and the new roots come into immediate contact with nutriment. If the ground in which the cuttings are to be planted is rather dry and of a loose nature, it is a good plan to puddle the cuttings before taking them to the place where they are to be planted. Puddling consists merely in mixing water with almost any good soil or clay, forming a composition of the consistency of thin mortar; the lower half or a little more of the cuttings are dipped in this, coating the part immersed in mud, which will adhere and prevent too rapid drying, as well as protect the cutting from injury by light and air during the operation of planting. Puddling cuttings, as well as the roots of plants, is practised extensively by gardeners and nurserymen as a ready means of giving temporary protection during transit from one part of the country to another, and also when transplanting in their own ground.

Where the climate will permit it, the proper time to plant cuttings is as soon as they are made in autumn, thereby avoiding all removals during the time they are forming a callus or producing roots; but in cold climates it is usually necessary to give protection of some kind during the winter by placing the cuttings in a position where the root-forming process may not be wholly suspended, even during the coldest weather. The cuttings of some kinds of trees and shrubs will grow freely under almost any treatment, while others require all the care that can possibly be bestowed upon them to insure the emission of roots and their future growth.

Success in growing such cuttings in the open air often depends as much upon the condition of the soil and the mode of planting as upon their proper selection and care during the preparatory stages. The soil in which cuttings are planted in the open ground should be deep, of a porous nature, and composed of materials that will absorb and retain a regular supply of moisture. The variation of climate should be attended with a corresponding variation in soil, which in warm latitudes should contain powerful absorbents, so that it will not become too dry in summer; while a soil for the same purpose in the more northern latitudes would be better without these absorbents. In this latitude a loamy soil of fine texture is perhaps the best; one that is not so fine as to cake and crack on the surface after heavy showers, or so loose that it will not retain sufficient moisture to supply the wants of the cuttings. The amount of moisture required by cuttings varies greatly in different species, some requiring little, while others a very large amount. The Poplars, Willows, Sycamore and many other kinds of trees and shrubs will grow readily, even if the lower end of the cuttings are immersed in water containing but very little nutriment. But, as a general rule, cuttings do not require more moisture than is held in suspension in well drained and friable soils.

Rank substances, such as undecayed vegetable and animal matter, should never be allowed near cuttings; and where it may be necessary to use manure, it should always be old, well decomposed, and thoroughly intermingled with the soil, and applied some time before the cuttings are planted.

Some propagators plant their cuttings and then cover the surface of the ground with manure, the juices of

PROPAGATION BY CUTTINGS.   149

which will percolate the soil at every shower and furnish nutriment in solution to the roots. I have found this to answer well in some seasons, while in others a fungus (Mushroom) would spread through the manure, and when it came in contact with the young growth on the cutting it was very likely to destroy it. This, however, can be prevented by frequently stirring the manure, and I may remark that frequent stirring the surface soil is almost equal to mulching for keeping the cutting-bed moist.

In some soils, and in hot and dry climates, covering the surface of the soil with coarse hay, straw, spent tan-bark or spent hops from the breweries, and other similar materials, will be of great service in keeping the soil moist and of an equable temperature.

MAKING CUTTINGS.—Nearly all plants emit roots more readily at or near their buds or joints than elsewhere; therefore, in making cuttings, it is always well to sever them just below a bud or buds, as shown in figure 49, thereby exposing the wood at a point from which roots appear to be produced most freely. It is true that some kinds of plants, like the Willow, Catalpa, and the common Quince, emit roots very readily from every part of the stem, and with these it is not necessary that they should be cut off below a bud. With some of the hollow stemmed plants, or those having a large pith, like the Sugar Cane and Bamboos among the large grasses, or the Mock Orange, Syringa (*Philadelphus*), and Deutzias among shrubs, the orifice is usually entirely closed, or nearly so, at the joints, as shown in figure 50, or opposite

Fig. 49.
CUTTING CUT JUST BELOW THE BUDS.

Fig. 50.
CUTTING OF HOLLOW STEM.

the buds on the young stems; and it must be apparent that there will be less liability of water lodging within the stem and causing decay if the cutting is severed just below the joint than at any other point.

The roots produced on a cutting are supposed to be formed from the sap or assimilated juices of the plant, deposited mainly between the bark and wood; but the stems and branches of some kinds of woody plants, having a large pith or hollow stem, will throw out roots from the inner portion, but this does not disprove the general theory of circulation of the sap, or that roots emanate only from the assimilated sap or nascent matter; because it is not impossible, nor contrary to the general principles of vegetable physiology, that a portion of the true sap of exogenous plants may not pass from the outer to the inner surface of the stems of young wood, at least when placed under artificial conditions.

There are some authors who contend that a bud, either latent or developed, is essential on a branch to enable it to produce roots; in other words, roots always proceed directly from a bud; and if a cutting is severed at a distance below a bud, the roots start from the lowermost one and push their way down under the bark and out at the end, establishing a communication with the source from which they are to derive their future nourishment. It was also claimed by some of the old vegetable physiologists, that in the same manner all the buds below the surface of the soil, which do not grow upright and form branches, produce roots by going down, overlapping and intermingling with those produced below them.

The erroneous theory of roots emanating only from buds doubtless originated from the fact that many kinds grow more readily from cuttings if they are cut off close to the base of a bud, as before stated, thereby strengthening the belief that at this point was located the materials from which roots are formed. To ascertain whether buds on ligneous plants must exist on or within a cutting to enable it to produce roots, it is only necessary to take a section of the stem between the buds (called the internode), of some kind of plant that has no latent buds—for instance, a young shoot of a Grapevine, or closely allied plant, and make such a cutting produce roots, which it will readily do if placed under favorable conditions. No buds will be produced or appear, yet roots will be produced more or less abundantly, or in proportion to the natural vitality of the cutting and the amount of available material which it contains. Of course the roots cannot grow to any considerable size, or for any considerable period of time, without the assistance of buds and leaves for assimilating the nutriments which the new roots absorb from surrounding materials.

That roots will live and continue to grow for longer or shorter periods, drawing sustenance from the material previously stored up in the plant, and upon that which they absorb without the assistance of leaves or buds, is too well known to be questioned. The tubers of herbaceous Pæonias and Dahlias, the thick fleshy roots of the common Rhubarb and many similar plants, will, when deprived of buds, live and continue to emit new roots for an entire season, and even for a longer period, but eventually perish, as they do not seem to possess the power of producing buds, except at the crown of the plant, where there is always a large number of buds.

The proper length to make cuttings will necessarily vary somewhat, according to the character of the plant from which they are made, as well as the manner of planting them. Cuttings of some kinds of trees that produce roots very freely, like the Willows and some of the Poplars, may be several feet in length, especially if set in moist or wet soils; while the other extreme in size may be a single bud with an inch or two of the branch attached. The single bud cutting is often employed with free-rooting kinds of trees, vines and shrubs of rare kinds, the cutting being laid down in a shallow trench, or set upright, and then covered to the depth of a half inch or a little more with light soil, or some material that will retain moisture well, and at the same time permit the young shoots to push readily through it. Single bud, or very short cuttings, when planted in the open ground, require more attention than longer ones, in order to prevent injury during dry weather; still, with proper care, they will usually make excellent and vigorous plants. As a rule, however, the most convenient lengths for what are called ripe-wood cuttings of deciduous trees and shrubs, are from six to ten inches, although they are often made much longer, which is not only a waste of material, but quite frequently a disadvantage when planting them; for in order to bury the cutting its entire length, as usually practised, it must either be set at an acute angle or the lower end placed almost beyond the influence of solar heat, an element as essential for the production and growth of roots on cuttings planted in the open air, as for the germination of seeds. Planting cuttings too deep should be avoided, as the farther from the surface they are the less solar heat

Fig. 51.—TRENCH FOR CUTTINGS.

they receive, and this is quite necessary to insure rapid growth, especially after they have become well rooted.

PLANTING THE CUTTINGS.—The surface of a cutting-bed should be level, smooth, free from lumps and stones. Draw a line across the bed and dig a trench, by placing the back of the spade against the line, pressing it down nearly perpendicularly; then throw out the soil to one side, making a trench, as shown in figure 51. Place the cuttings against the perpendicular side of the trench, as

PROPAGATION BY CUTTINGS. 153

shown, and two to six inches apart—according to the size and kind of cutting planted—and the upper end an inch or two above the surface of the soil. Draw in a little soil and press it down firmly with the foot, or with a pounder made from a piece of two-inch plank, shown in figure 52. After the soil has been packed firmly about the base of the cutting, the trench may be filled up level with the surrounding surface. With many kinds of plants the packing or firming of the soil around the lower end of the cuttings is a very essential point, and often the whole secret of success. This is particularly so with those kinds that produce roots mainly from the lower end, where the wood is exposed to the soil. It is true

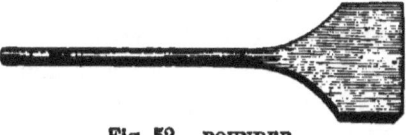

Fig. 52.—POUNDER.

that cuttings of some kinds of trees and shrubs will grow readily if merely thrust into the ground, almost anywhere, and at any season; but to raise such plants requires neither experience nor skill; consequently they only call for a passing notice.

The distance between the rows of cuttings will depend entirely upon the manner in which they are to be cultivated. If a cultivator or plow is to be used, then the rows should be two and a half or three feet apart; but if to be cultivated entirely with the hoe, one and a half to two feet wide will be sufficient. In some soils and climates, the plants raised from cuttings may be left in the ground where grown until the following spring, and then transplanted; but in cold climates it is best to take them up in the fall, and store them in some convenient, safe place during the winter.

In making and planting cuttings of the ripe or dormant wood of evergreen plants, the same general plan should be pursued as with deciduous plants, except that the cuttings are not usually made as long, and they must not be entirely excluded from the light, or wholly buried in the soil; for in making them, the leaves are left on that part of the cutting which remains above ground when planted. If covered entirely, as we do with deciduous cuttings, they would soon decay. The leaves of our hardy evergreen coniferous plants are covered with a compact epidermis, which does not permit them to either absorb or exhale moisture very rapidly when in a dormant state, and this peculiarity in structure admits of their being placed in such a position that roots will form while their leaves do not suffer. In cold climates they should be placed in what are termed cold frames, and covered with glass that has been dimmed with some kind of a wash that will prevent the direct rays of the sun reaching them, and still admit sufficient light to keep them in health. In the coldest weather the frames may be covered with straw mats to keep out the cold and prevent severe freezing. The Arbor-vitæs, Evergreen Box, Yews and Junipers, are quite readily propagated from cuttings made of the dormant wood in the fall, although this mode of propagating these trees is seldom practised, probably because there are others less hazardous, and perhaps more convenient.

CHAPTER XII.

BY CUTTINGS OF IMMATURE GROWTHS.

When propagating by cuttings of the young, growing wood, or succulent parts of herbaceous plants, we are operating with an active vegetation instead of one that is dormant. In the cuttings made from ripe wood, there is a supply of organized material from which roots are produced; but in those made from the young and growing parts of plants, this is only in a state of transmutation, and we aim to continue or accelerate the change, and not to check it by our operations. To accomplish this, it is generally necessary to surround the cutting with a warm, moist, and somewhat confined atmosphere, so that the exhalation, which is very rapid in an open situation, can be controlled. If the quantity of moisture given off by the leaves is greater than that absorbed, then they will surely droop, and artificial applications will be required to revive them.

Various devices are employed by propagators to secure that peculiar state or condition of the atmosphere most suitable to the growth of cuttings of different kinds of plants. Glass is the principal material used for covering propagating houses, because it is the most durable, and through it plants receive light, which is indispensable to the growth of all those belonging to the higher orders; while, at the same time, heat and moisture are under the control of the propagator. These houses may be made of almost any size or form to suit the convenience or taste of the owner, but those with a double pitch or span roof, as shown in figure 53, are the most common, and probably the most convenient for heating and ventilating, as well as exposure to the light and direct rays of the sun during the greater part of the day.

The illustration on page 156 shows the interior perspective view of a section of a well-arranged propagating house. It may be made of almost any desired length or width, but eighteen or twenty feet is about the usual width of a span-roof house of this kind.

The side walls of the house may be of brick, stone, concrete, or other durable material, and in very cold climates it is well to bank them up on the outside with earth, and sod over the embankment. The house should stand with ends north and south, although a slight deviation from this direction will not be a serious objection. The side walls should be low, not more than four or

156 PROPAGATION OF PLANTS.

Fig. 53.—SPAN-ROOF HOUSE.

five feet high, and a row of low frames may be placed under the eaves on the sides, as shown in figure 53, or be omitted, according to the taste of the builder or nature of the plants to be propagated.

The glass should be of the best quality, and double thickness is preferable to the single. Embed the glass with putty, and fasten with glaziers' tins, but put no putty on the outside ; use nothing but thick white lead paint in the joints between glass and sash. The size of glass is immaterial, but if the best and heaviest is used, the panes may be of any size, from seven by nine up to ten by sixteen, or even larger. Large sized glass, however, is more expensive than small ; it breaks quite as readily ; consequently, repairs, where large sized panes are used, are likely to be the most costly. The furnace and potting room should be placed at the north end of the house, if such an arrangement can be made without inconvenience, and then the south end may be of glazed sash, as shown in figure 53.

The best and most economical mode of heating a large house is by hot water. For this purpose there are several kinds of boilers in market, each of which has its friends among the florists and nurserymen, but all of those now in common use are economical, safe, and generally give satisfaction. The hot-water pipes should lie side by side, although they are sometimes placed one over the other, when it is inconvenient, for want of space, to place them the other way. The flow-pipe passes under one of the side frames, thence through the back under the middle one, and then under the frame on the opposite side. The return-pipe passes back along by its side, both lying on iron rests made for the purpose. This arrangement gives eight pipes the whole length of the house, besides the elbows and the few feet that it takes to cross the end. The center frame has four pipes under it, while those on the sides have but two. In mild climates, where less artificial heat is required in winter, the pipes may pass only under the two outside frames, and the center one may be used for the plants after they have become well rooted, or for other purposes. For convenience in reaching the plants, and economy of space, the center stage or frame is made double the width of the side frames, and the depth of all depends upon the purposes for which they are to be used. If for small cuttings only, then they may be quite shallow—not more than six to eight inches deep; but if grafting is to be done, then a greater depth will be required, or from one to two feet in depth; but if the house is to be a kind of "general utility house" in propagating plants, then the frames may be made of various sizes and depths. The center frames may be deep enough to hold potted stocks of good size, and the side frames must be shallow, for growing small cuttings. The pipes under the frames should be entirely shut in, so that the greatest heat in the house will be under the frames, in order to give what is termed "bottom heat" to the cuttings, exciting root-growth in preference to growth of the leaf and stems. There should be small doors placed along the entire length, opening into the passage-ways, that may be opened to let the heat escape into the house when necessary to raise its temperature, or to lower that under the frames. The passage-ways between the beds should not be less than two and a half feet wide, and three will be better. If the frames are over four feet wide it will be inconvenient to reach across them, and a house eighteen feet wide will allow of three rows of frames, and two passage-ways of three feet each; or the outside frames may be three feet wide, and the center one—as it can be reached from both sides—may be six feet.

Arrangements for ventilating the house may be made to suit one's convenience or fancy, but the openings should be mainly at the top or near the peak of the roof.

The slope of the roof should be at an angle of from thirty-five to forty-five degrees, varying somewhat according to latitude, although water, snow and ice will pass off more readily from a steep roof than a flat one. A single roof, or lean-to, propagating house is perhaps a little more economical than a span roof in very cold climates; being less exposed to cold winds, it takes less fuel to heat it. Its construction and interior arrangement may be the same as the span roof, only the center frame, or table, is divided lengthways by the main or back wall, which should run east and west, in order to have the roof slope to the south. The lean-to house may be just one half the width of the span roof, or a little wider if necessary, and the furnace-room, potting and storing sheds, should extend the entire length of the back wall as a protection.

While professional gardeners depend mainly on a regularly constructed propagating house for multiplying those

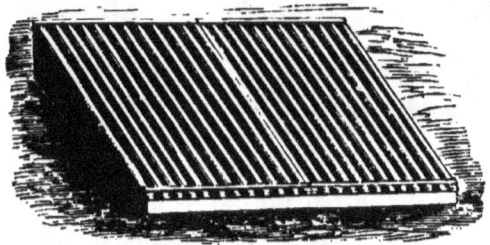

Fig. 54.—LATH SCREENS.

plants requiring artificial heat to force them to produce roots, they often employ other means and structures, such as the ordinary hot-bed, where a high temperature and bottom heat are secured by the use of stable manure, leaves from the woods, spent hops, and other kinds of fresh vegetable substances, which generate heat during fermentation. In summer the ordinary hot-bed frames may be employed without bottom heat for multiplying many kinds of plants from green cuttings, especially those known to produce roots freely and with little care, when placed in a confined, moist, and rather warm atmosphere. When grown in frames in the open ground, the cuttings will need to be shaded the same as when planted in the house, and this may be given by using lath screens, as shown in figure 54, or the glass may be dimmed by applying with a brush a wash made of skim milk with common whiting. The cuttings in frames will need attention when growing, for water must be regularly supplied, and air admitted occasionally to prevent the temperature from reaching too high a point during warm, bright days. Common white muslin, made water-proof, or coated with boiled linseed oil and well dried before

 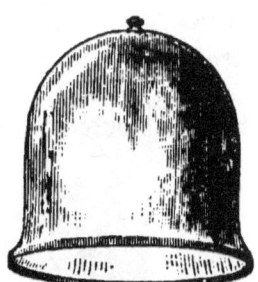

Fig. 55.—HAND GLASS.   Fig 56.—BELL GLASS.

using, may be employed for covering the frames in which cuttings are rooted during the summer months, but glass is preferable, and as it is so much more durable, it is in the end the cheapest.

What are called hand glasses (figure 55), and bell glasses (figure 56), may often be employed with advantage, where only a few cuttings are to be raised, either in the open ground or in an ordinary greenhouse. These glasses are also useful in covering delicate seeds and the spores of ferns, which, owing to their minute size and fragile structure, cannot safely be buried in ordinary soil.

IN WHAT MATERIALS TO PLANT CUTTINGS.—Various kinds of materials have been recommended, and are still used, in which to root cuttings made of the green and immature parts of plants. Sand, burnt clay or pulverized brick, charcoal, peat, Sphagnum moss, and various combinations of these and other materials, have been employed by propagators; and while any one, or all of them, may sometimes answer the purpose, still, the more delicate the cutting, and difficult the plant is to propagate in this way, the purer must be the material in which the cuttings are set to insure the production of roots. In a previous chapter I have shown that the exposed cells of a branch or leaf taken from certain plants will imbibe various poisons and colored liquids presented to them, which are successfully resisted by the roots of the same plant; consequently, the material best and safest to use in growing green cuttings in general is one free from all salts, metallic oxides, humus or coloring matter of any kind. The most abundant and available material of this nature is pure silex (sand). That which may be obtained from the banks of fresh water streams, ponds and lakes is generally the purest and best, but that found in the ordinary sand-pit may be used with safety, if well washed to remove the oxides of iron and other metals which are usually present, giving to the sand its red, yellow, or other shades of color. When ordinary building or pit-sand is to be used for propagating plants, it should be dug out and exposed to the weather for a few weeks or months previous, in order to hasten the further oxidizing of the metals which it may contain, and facilitate their removal by washing. Sea-sand is, as a rule, too fine for such purposes, besides it contains various salts, and frequently magnetic iron in large quantities. Many a gardener has sustained serious losses from attempting to propagate certain kinds of plants, through negligence in this matter of obtaining pure sand in which to place his cuttings.

If pure sand is placed in pots, boxes or frames, with

sufficient openings in the bottoms to allow of rapid drainage, there is little danger of applying too much water on the surface, as the sand will only retain a certain amount; all in excess passing rapidly away. The roots being produced from the organized matter already in the cutting, no further nutriment than that contained in the water applied, is required until the cutting is well supplied with roots, when it should be removed from the sand and potted in good soil, with which more or less sand is intermingled, varying the amount according to

Fig. 57.—CUTTING OF GREEN WOOD.

the nature of the plant being propagated. Plants of a soft, succulent nature, like the Coleuses, Begonias and Geraniums, should at first be placed in a rather light and porous soil, while a more firm one may be given to ligneous plants of firm texture; although it is better to err, if at all, in placing the newly rooted cuttings in rather too light a soil than one that is too heavy.

PREPARING THE CUTTINGS.—In making cuttings of the young or succulent shoots of plants, a portion of the leaves, should be allowed to remain on the cutting, for the purpose of assisting in the assimilation of the sap and preparation of the material required to produce roots. In selecting cuttings of ligneous plants, it is advisable, when convenient, to take the small, short side shoots that may be cut off close to the more mature wood, leaving the hip or ring of half ripened wood attached at the base; for, as I have stated on a previous page, in regard to ripe wood cuttings, there is usually an aggregation of buds, and more available organized matter, at this point than elsewhere. The lower leaves on the cutting should be removed. This should be done with a sharp knife, and the wound made left with a smooth surface; for severing with a dull knife, or crushing with shears, and even pulling off the leaves by hand (as often practised by careless gardeners, who are in such haste that they seldom do their work well), will not answer when making cuttings of the soft shoots of delicate plants. The old saying that anything that is worth doing at all, is worth doing well, is certainly applicable here, for no one can be too careful in making and planting cuttings of the green shoots of woody plants and succulent stems of herbaceous plants. It is also well to make up such cuttings in the house, or where they will not be exposed to the direct rays of the sun while they are being made.

It is not always practicable to make cuttings of this form, for some woody plants will produce few or no lateral shoots, unless the leading ones are pinched back for the purpose of forcing out side branches, and where the latter are not available the terminal, or leading shoots, may be used for cuttings, and with some kinds of plants the entire new growth of the season may be divided up for this purpose. In some cases the half ripened, or nearly mature shoots of herbaceous plants, are better for cuttings than a younger growth. With the half woody and herbaceous plants, the tips of both upright and lateral shoots are used for cuttings, and in making them the lower leaves are cut off, as shown in figure 57. The cuttings should be cut square across, just below a bud, or the axil of a leaf, and of twigs firm enough to be severed without breaking or crushing under the blade of a sharp implement. In making very small and slender cuttings, like those of the Heath (*Ericas*), a razor in good

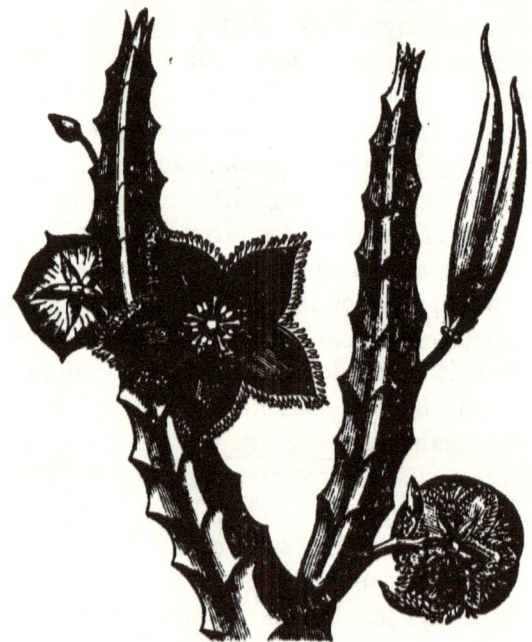

Fig. 58.—STAPELIA GLAUCA.

order is none too sharp an implement for this work, and the heel of the cutting may be placed on the thumb nail or a piece of horn when being smoothed off before planting.

No general rule can be given as to the exact time for separating the cutting from the parent plant; with some it is better to take them off while the growth is very young and tender, while with others it should be somewhat mature and firm. Cuttings of the succulent plants, like the Stapelias (figure 58), and the Cactuses and others of a similar structure, are benefited by a few hours or even days of drying before planting, or they may be set in dry sand or soil a few days before any water is applied. But at no time should cuttings of these plants be watered very freely, but the sand in which they are set should be kept only slightly moist. Some propagators make a practice of wilting the cuttings of Geraniums, Acacias, Banksias, and many other kinds of plants, before placing them in the cutting boxes or frames, but others appear to produce just as good results without it, planting the cuttings as soon as made.

CUTTINGS OF THE LEAVES.—The leaves of many kinds of plants may be employed in making cuttings whenever it is necessary for their rapid multiplication. In propagating woody plants by cuttings of the leaves, the leaf is usually taken off entire, with the petiole or leaf-stalk attached, as shown in figure 59, the leaf-stalk in this case representing the stem of the ordinary cutting. Such cuttings should be taken while the leaf is fresh and in a condition for supplying the proper materials required for the production of roots and buds, and placed in a warm, moist and confined atmosphere, or treated in the same way as the ordinary green cutting. Not only can the leaves of the common Lilacs, Roses, and various other kinds of hardy shrubs and trees be propagated by cuttings of the leaves, but there are hundreds of the common and

Fig. 59.—LEAF OF LILAC.

rare species and varieties of tender greenhouse and bedding plants that may be, and some often are, multiplied in this manner. Certain kinds produce roots so freely from their leaves that it is not necessary to use or preserve the leaf-stalk, but the leaf may be laid down with its underside in contact with the sand, and little wooden pegs thrust through it as shown in the Begonia leaf,

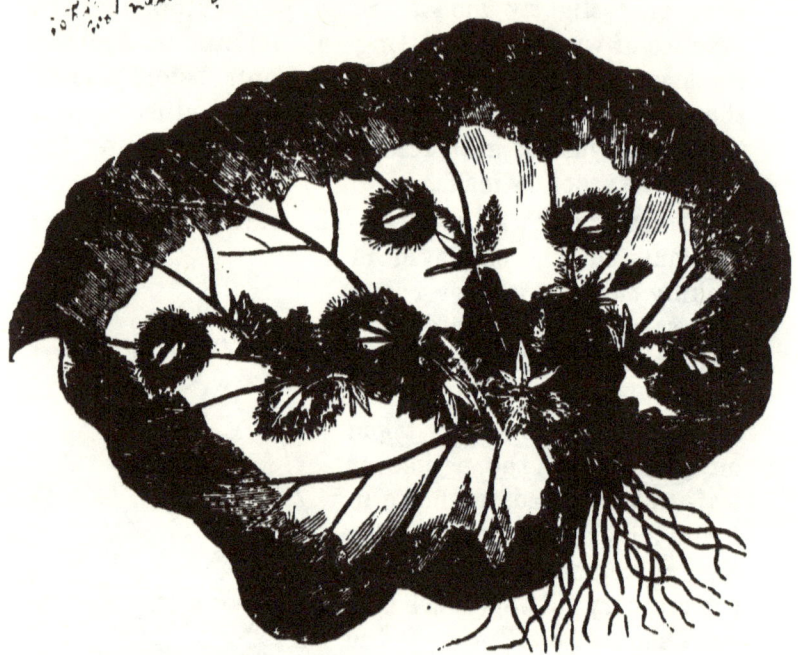

Fig. 60.—LEAF OF BEGONIA, PRODUCING YOUNG PLANTS.

figure 60, or the leaf may be cut up into small pieces and then set upright in sand, when each part of the leaf will soon produce roots, buds, and then new leaves, as shown in figure 61.

The thick, fleshy leaves of the common *Bryophyllum calycinum*, as they fall to the ground, will often throw out roots and buds from their edges, and, if left undisturbed, these buds soon become vigorous individual plants. Not only are the *Bryophyllums, Begonias, Gesnerias*, and many other kinds of closely allied plants, propagated by cuttings of their leaves, but among the succulent Cactuses, Stapelias, Echeverias and Cacalias, either their leaves, or leaf-like stems, are generally employed in multiplying the various species and varieties in cultivation.

SETTING THE CUTTINGS.—The depth and distance apart at which cuttings should be set must necessarily vary according to their length and thickness, but, as a rule, one-half their entire length should be covered when set in a position for growth.

The more usual lengths for green cuttings are from two to four inches; consequently, the sand in the boxes need not be of greater depth than to give room for the production of roots, and insure the retention of sufficient amount of moisture around the base of the cutting. If the sand in the boxes is made rather moist, a sharp or pointed stick, or dibber, may be employed for making the small holes in it for the reception of the cuttings, but where the cuttings are small, and a large number are to be set, a pane of glass is a better implement, for by pressing one edge down into the sand, a straight, narrow trench is quickly made across the box, and into this the cuttings may be set very rapidly, and at any distance apart desired. When a row has been set, the glass may be inserted near it, and with a slight side movement the sand pressed firmly against the cuttings, or the sand, or other material, may be pressed down around each cutting, separately, with the fingers or the

Fig. 61.
PIECE OF BEGONIA LEAF,
USED AS A CUTTING.

large end of the dibber. After filling the box, the cuttings may be sprinkled with water, which will aid further in compacting the sand about the base of the cuttings.

The professional florist, with the conveniences for propagating plants in quantities, will seldom have occasion to employ anything of less size than boxes holding several dozen or a hundred cuttings, but the amateur may often have occasion to raise a less number of plants of certain species or varieties, and in doing so have recourse to ordinary flower-pots for this purpose, placing them in a window, or by the side of some building or board fence, where the requisite amount of heat may be obtained during the summer months.

When green cuttings are set in pots filled with sand, and placed in the window of an ordinary room in summer, or even plunged in a half-shady place in the open ground, they are very likely to be occasionally neglected, hence suffer for want of moisture. To prevent this, and insure a constant and regular supply, the device shown in figure 62 is frequently employed. In this two pots are used, one inside of the other, with sufficient difference in their sizes to leave a space between, which is filled with sand, *c c*, and into this the cuttings are inserted. The hole in the bottom of the inner or smaller pot is stopped with a cork, and the pot is then employed as a reservoir for holding water, *a*. If the smaller pot is of the ordinary unglazed kind, enough water will percolate through the sides and bottom to keep the sand fairly moist, but never saturated. The small, inner pot should be kept filled with water, or at least not be allowed

Fig. 62.—DOUBLE POT FOR CUTTINGS.

to become entirely empty. By setting the cuttings close to the rim of the inside pot, a bell glass may be used for covering the cuttings, thereby insuring a close, moist atmosphere, and preventing too rapid evaporation of the juices of the cuttings through their leaves. By employing larger pots and filling them about half full of sand, the cuttings may be covered with a large pane of window glass, laid flat on the top of the pot. Boxes six to eight inches in depth may be employed in a similar manner, and often with excellent results, for cuttings of Geraniums, Coleuses, Fuschias, and other kinds of plants that are readily propagated from green cuttings. With plants that are not so easily propagated, a similar arrangement of the pots may be employed, and the space between the two filled with moss, tan, or even sand, and the center one also filled with sand, placing the cuttings in this, instead of around it, as in the preceding arrangement, employing two bell glasses, as shown in figure 63—a, the larger bell glass; b, the inner, or smaller one; c, cutting; d, sand in the inner pot; e, filled space between the pots; f, the outer, or larger pot. With this device, and proper attention for securing a temperature of seventy to eighty degrees Fahrenheit, cuttings of many kinds of plants usually considered quite difficult to propagate, may be forced to produce roots in a few days or weeks.

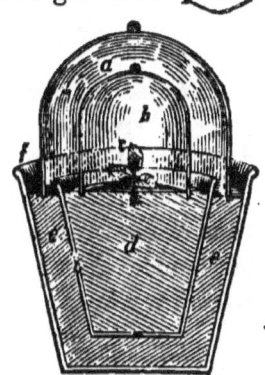

Fig. 63.—DOUBLE BELL GLASS FOR CUTTINGS.

Many other devices are employed in propagating plants from green cuttings, but the principles governing the operation are the same in all, and while the propagator's ingenuity or fancy may lead him to vary the mode of doing a thing, he seeks only the rapid production of roots.

CHAPTER XIII.

PROPAGATING BY LAYERS.

The same principles given for propagating plants by cuttings are applicable to layers, with this difference: layers of the green and growing parts do not need to be surrounded with a confined atmosphere, in order to prevent a too rapid evaporation of their juices through the leaves. A layer is only a cutting that is allowed to remain attached to the parent plant until it has produced roots through which it may collect sustenance for self-support.

Various methods are employed to produce this result, such as ringing, girdling, twisting, tonguing, or partly dividing that portion of the stem or branch on which it is desired that roots shall be formed. All these distortions of the stem or branches of the plant layered are for one object, that is, to check the downward flow of sap. Roots then become necessary for supplying sustenance to the cutting, or layer, and are consequently formed. The most common method of preparing layers is that of making a tongue on the under side of the branch. The operation is performed thus: Make an incision in the branch or part of the plant to be layered, just below a bud, cutting through the bark and into the branch to the depth of one-quarter to one-half its diameter; then pass the knife upward for an inch or more, according to the size and nature of the plant being layered, splitting the branch lengthways, forming the tongue as shown in figure 64 at *a*. The branch is then bent down and fastened in its place by means of a hooked peg, *c*, and the end tied up to a stake, *b*, as shown in figure 64. That part on which the incision is made is covered with soil or other material that will exclude it

PROPAGATING BY LAYERS. 171

from light and air, while at the same time keeping it moist, thus aiding the development of roots. In making layers of certain kinds of small herbaceous plants and slender vines, it will not be necessary to use pegs or stakes to hold the layer in place; but with larger plants, they are usually needed for keeping the layered branch steady and in one position, while the new roots are being emitted.

Twisting, coiling, or notching the branch, so as to partly separate the fibres, will often answer the purpose

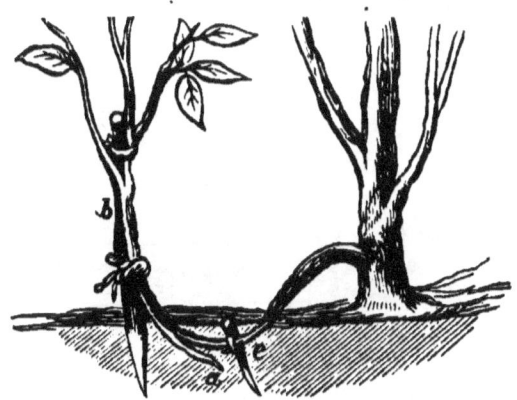

Fig. 64.—LAYERED BRANCH OF A TREE.

equally as well as tonguing. Girdling the branch to be layered, by removing a ring of bark a half inch or more in width, is another mode of exposing the alburnum to the soil and checking the flow of sap in woody plants. Bending the branch at a very acute angle will often answer the same purpose, but this can only be done with safety on plants with tough wood and bark. There are many kinds of ligneous as well as herbaceous plants that will produce roots freely, and in a few days' time, from branches merely laid on the surface of the ground and covered with almost any kind of material that will keep them moist; but there are others that produce roots very tardily, or not at all, even under the most careful and skilful manipulation.

In preparing layers of those plants which produce roots very slowly, it is best to distort that part which is buried as much as possible with safety; but with many kinds no disturbance of the natural condition is necessary. With nearly all kinds of climbing and trailing plants, such as Wistarias, Tecomas, Honeysuckles (*Lonicera*), Grape, Passifloras, etc., it is only necessary to lay the stems in a shallow trench, early in spring, and when the buds push into growth, draw the soil back into place; thus treated,

Fig. 65.—LAYER OF VINE.

each shoot will produce a plant. Roots will usually be emitted in abundance the entire length of the old stem, as shown in figure 65.

When roots are not produced as freely as desired, then the layered cane may be bent, as shown in figure 66, a method quite generally practised with the Wistarias, Passifloras, and similar vines, especially when a large number of plants are sought, in preference to a few of more vigorous growth. If only a few strong plants are wanted, then the cane may be layered with one single bend, as practised with shrubs and trees. (See figure 64.) In preparing shrubs and trees for the purpose of producing layers therefrom, it is often necessary to head them back severely, the year previous, so that a large number of shoots shall start from near the ground. Plants thus prepared are technically termed stools; and if all the new shoots are layered in any one sea-

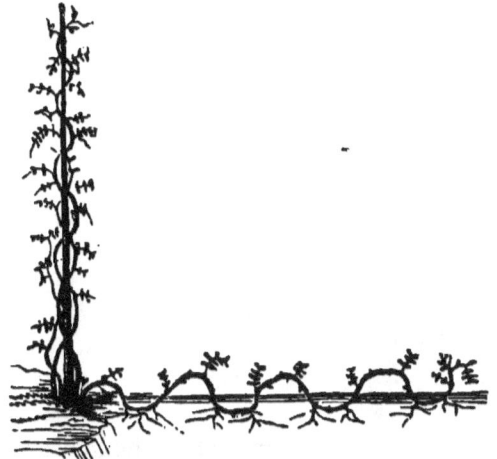

Fig. 66.—BENT LAYER OF VINE.

son, then none should be layered the next, but the new growths be allowed to grow unchecked, so that the parent stock may regain that strength of which it has been deprived by excessive layering. With some kinds

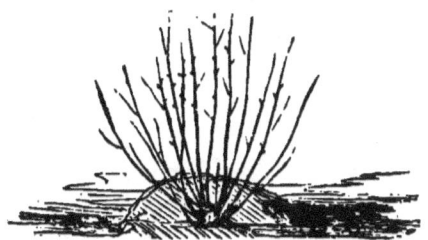

Fig. 67.—STOOL BANKED UP.

of plants it is not necessary to bend down or layer the shoots, but merely to bank them up with soil, as shown in figure 67, and when the shoots have thrown out roots from their base, they are slipped off from the main stalk and planted out separately. The Quince, Hibiscus, Cercis, Paradise Apple, and many other kinds of trees and shrubs that produce roots readily, are extensively propagated in this manner. It is certainly more convenient for the purpose of layering, if the branch is near the ground; yet it is not positively necessary, because soil, or other material which will answer the same purpose, may be elevated to the branch to be layered.

Fig. 68.—LAYER IN POT.

A platform may be made, upon which to place the soil (figure 68), or boxes and pots filled with it may be suspended among the branches of a tree and the shoots layered therein. But in making layers of this kind, it will require more care to keep the soil moist around the layer than when they are placed in the earth in the usual manner. It will seldom be necessary to make layers of branches that cannot be bent down to the ground. Still, instances may occur, where it will be desirable to secure some freak in this way.

TIME FOR MAKING LAYERS.—The proper time, or season, for making layers is as variable as that for making cuttings. But, as a rule, layers should be made while the parent plant is growing most rapidly, because roots will be produced at such times more readily than at any other, although with several kinds it will make very little difference, as they produce roots freely under almost all conditions and from all parts of the plant. With the larger proportion of both deciduous and evergreen trees and shrubs, layering should not begin until the leaves have fully expanded, and the new growth of the season is fairly under way. If layered earlier, many of the deciduous trees and shrubs will "bleed," as it is termed, from the wounds made on the layered parts, and the sap flowing from these wounds will often corrode and otherwise injure the exposed cells and entirely prevent the production of roots therefrom. Wounds made in the branches of coniferous trees during the winter and early spring months are usually soon covered by the exuding resin, the severed and otherwise exposed cells thereby becoming fully protected, not only against the influence of moisture from without, but it effectually prevents the formation of a callus and production of roots. For this reason, such conifers as Pines, Spruce, and Firs should always be layered at a time when the sap is thinnest and flowing most rapidly, as during the first growth of spring and early summer. With some kinds of hardy deciduous trees and shrubs the autumn is the better season in which to make layers, for the reasons given for making cuttings at that season, but such variations will be noted further on. The advantages and disadvantages in propagating plants by layers are:

First, It is a very simple process of multiplying plants by which an inexperienced person may propagate many species and varieties that require greater skill by different processes; in other words, it is especially valuable to amateurs who possess neither skill nor facilities for propagating plants by the more advanced and superior modes.

Second, It is a certain mode, with by far the larger number of ligneous species, as the parent plant sustains the layer until it has produced roots through which it may, when transplanted, derive nutriment for self-support; but the roots produced on layers are seldom of as firm a texture, neither do they mature as early as these produced on cuttings: hence their inferiority.

Third, With many kinds of plants much larger specimens may be obtained in less time than is possible by cuttings, and, as size is frequently more highly appreciated than quality, large layers are usually in demand.

CHAPTER XIV.

PROPAGATION BY SUCKERS AND DIVISIONS.

Suckers are shoots which are produced from latent or adventitious buds on the subterranean stems or roots. Shoots that spring up from the base of the main stem of a plant are often referred to under the name of stem-suckers, but "sprout" is by far the more appropriate name, as they are of a very different origin, and should not be confounded with the true suckers of plants. There are hundreds of different kinds and species of plants which, under certain conditions, produce sprouts quite freely from their crowns or the base of their stems, but seldom or never a true sucker from the roots.

The common Dahlia and Pæonia are familiar tuberous rooted plants that produce sprouts or stems from their crowns, but no buds or suckers from any part of the tuber or root below it. Among trees, the Catalpas, Maples, Magnolias and Chestnuts are well known kinds that produce sprouts freely, but seldom or never a sucker.

To increase by suckers appears to be one of nature's methods of multiplying certain kinds of plants, and when these are brought under cultivation it becomes a ready and simple means of propagation. It is very probable, however, that long continued propagation by suckers will intensify the habit until the plant multiplies so rapidly in this way, that it will lose much of its original value, even if it does not become an intolerable nuisance to the cultivator. The propagation of plants by natural suckers is certainly a convenient process, but upon the whole it tends more to the degeneration of the species or variety thus multiplied than to improving it; consequently, it should never be resorted to except with some of the simple species, like the Raspberry and Blackberry, or herbaceous plants, which are frequently transplanted or renewed from seed.

Suckers from old Pear, Cherry, Plum, and other kinds of fruit trees may, under some circumstances, be worth preserving, especially when other means of propagation are not at hand; but trees raised from these suckers are seldom as vigorous, productive and healthy as those propagated by other and more scientific methods. The same is true of many kinds of forest trees which sucker more or less freely; and while there may be no difficulty in transplanting the suckers and making them grow, they are, as a rule, inferior to seedlings.

There are many kinds of herbaceous plants, such as Japan Anemones, the Pachysandras, Lilies-of-the-Valley, and some of the herbaceous Spiræas, that are usually propagated by suckers, but they are so frequently transplanted and divided that no apparent deterioration has as yet been observed from this long continued mode of propagation.

PROPAGATION BY DIVISION.—In propagating plants by divisions, we take advantage of all the various forms of growth wherein buds or sprouts appear on the crowns, side shoots, or the base, of the stems. With many kinds of herbaceous plants, division of the clumps of the buds, stems and roots is the most common mode of propagation. There are also various kinds of ligneous plants which may be rapidly multiplied by divisions, employing for this purpose the sprouts which spring up from the base of their stems. The number of these sprouts may be greatly increased by severely heading-back the main stems early in spring, thereby forcing the stump to produce a large number, and permitting the entire forces of the roots to be expended in producing buds and sprouts. After these sprouts have reached a moderate size they may be banked up and their bases covered with soil, to induce the formation of roots at this point, as described in the preceding chapter; but this banking must be varied according to the nature of the plant operated upon; for, in some instances, to place earth around young, succulent sprouts would destroy them, while on others, roots would be produced more freely than if the operation was delayed until they were nearly or quite mature. When such sprouts have become well furnished with roots, whether it be at the end of the first, second or third season, they may be slipped off and planted out separately. In propagating plants by divisions, the best time for separating sprouts from the parent stock, or dividing up clumps of herbaceous plants, is when they are in a dormant state; but this does not occur in all the plants indigenous to any country, at one time, or during any one season. It is perhaps true that the larger proportion of the plants cultivated in the gardens of temperate climates are in a semi-dormant condition during the colder months; still it is well known that the roots of certain kinds grow very rapidly, while the aerial parts are to all outward appearances in a state of complete rest.

With by far the larger number of the species of plants cultivated in cold climates, spring is the best season for dividing and transplanting, because the warm rains which usually fall at this season favor the production of new roots as well as buds and growth of stem. But in mild climates, where the ground does not freeze, or at least not to any considerable depth in winter, the autumn months would be preferable, in which to do such work, to those of spring.

There are, however, many kinds of bulbous, tuberous, and fibrous rooted herbaceous plants indigenous to cool and cold climates that make all their stem and leaf-growths early in spring and then rest during the warmest weather in summer. No amount of rain or degree of heat will force them to make a second growth. With such plants, among the most familiar of which are the spring flowering bulbs—the Narcissus, Hyacinth, Tulip, etc., or the fibrous and tuberous species of early Anemones, Spiræas, Dicentras and Perennial Poppy—dividing the bulbs and clumps and transplanting should immediately follow the ripening of the foliage, for this time is the beginning of the dormant season with these plants.

Root-growth and development of flower-buds will take place later in the season, and while no stems may appear above the surface, they are gathering materials for the ensuing season's growth. In a few instances, as in the common Garden Lily (*Lilium candidum*) and Oriental Poppy (*Papaver orientale*) a part of the main leaves of the plants push to the surface early in autumn. Owing to the wide difference in the habits and structure of plants, no one season, or time of year, can be considered as best for dividing and transplanting all kinds.

In the common herbaceous species of the Pæonia and Dahlia we have types of a class of plants that produce tubers with buds clustered at the apex or crown, and not distributed over the entire surface, as seen in the Potato and American (commonly called Jerusalem) Artichoke. In propagating the varieties of either class, we have only to separate the buds, leaving a tuber or a small portion of one attached to the bud, for the purpose of supplying it with sustenance until it emits new roots and tubers. Many of the fibrous-rooted plants which grow in tufts or stools, as seen in some species of Grass, Garden Pinks, and other common plants, are usually propagated by divisions. All plants which naturally produce buds, bulbs, or tubers on their roots or subterraneous stems, are usually more readily propagated by root divisions than are those which do not show such formations, but, as I have already shown, with many kinds the existence of buds is of no importance whatever, because there is an inherent power in the plant sufficient to produce them whenever they are required.

PROPAGATION BY RUNNERS, OR STOLONS.—There are many species of plants, such as the common Strawberry, some of the Potentillas, Saxifrages, and various species of Grasses, which throw out long, slender shoots, producing buds, leaves and roots at the nodes or joints. Plants of this kind require very little artificial aid in their propagation; still we can hasten development and growth by keeping the surface of the ground rich and loose, applying water whenever necessary. In propagating the common Strawberry by runners, the young plants may be layered in small pots filled with rich soil, and then plunged in the bed and in the most convenient position to receive the undeveloped plant when ready for producing roots or layering. By this mode the roots are confined within the pot, forming a close, compact ball, which facilitates transplanting and prevents any serious check to the growth of the young plant.

CHAPTER XV.

PROPAGATION BY ROOT-CUTTINGS.

The propagation of plants by root-cuttings has its advantages and its disadvantages. With some kinds, such as certain species of the Raspberry and Blackberry, it will yield plants far superior to those usually obtained in the form of natural suckers; but with fruit trees, such as the Pear, Plum, and Cherry, it should never be employed, except when no other mode can be made available, or for the purpose of saving some choice variety from becoming wholly extinct. Trees that naturally produce suckers are likely to have this habit intensified by continued propagation by either suckers or root-cuttings; but there are others upon whom it does not appear to have this effect, and with such the propagation by root-cuttings is to be recommended as a very cheap and excellent mode of increasing the number.

Plants that produce suckers freely can usually be grown from root-cuttings; but we are not limited to these alone, for there are many kinds that seldom or never produce a sucker under natural conditions, that under artificial conditions may be readily propagated by root-cuttings. Why one plant possesses so much vitality that every portion, from the leaves to the most minute point of a root, can readily be made to produce a separate plant, while with another every attempt at such division is unsuccessful, cannot be scientifically explained. The fact is apparent—the cause unknown.

Root-cuttings of hardy plants are usually made in autumn, for the same reasons given for making cuttings of ripe wood at that season, and in all such operations it is well to keep in mind that it takes time for changes and the reorganization of cellular matter. The size of the cutting will necessarily depend somewhat upon the kind of plant under propagation, and the manner of keeping through the winter must also be varied considerably, owing to the difference in the nature of the roots. Some kinds will require more warmth and moisture to develop buds and new roots than others; but the propagator who closely watches the progress of his work can readily vary the conditions to meet the requirements of each and every kind. He can hasten the root-forming process by giving more heat, or retard it by ventilation and lowering the temperature.

As a packing material in which to store root-cuttings of all hardy plants, none is equal to the common Peat-moss, or *Sphagnum* from the swamps. It will hold moisture well, is of a cool nature, and will not ferment and heat, even when packed closely and subjected to a high temperature. It is also a clean and sweet material in which to pack away cions, roots or cuttings of any kind, whether they are to be preserved for a few days or for months.

The next best material for this purpose is clean, sharp sand, and it is more generally used, probably, because very abundant, and less labor is required in packing away as well as removing the cutting, than when moss is used. Ordinary soil, if of a rather light, sandy nature, will answer very well as a material in which to store cuttings, but either moss or sand is better.

In making cuttings of the roots of such hardy plants as the Raspberry, Blackberry, Juneberry, and various ornamental shrubs and trees that are usually propagated by this mode, the roots should be taken up as soon as the plants have ceased growing in the fall, and cut into pieces of from one to three inches in length, but in some coarse-growing kinds, such as the Paulownia, they may be a little longer. Then prepare some boxes, by boring holes in the bottom to insure good drainage, and place a layer of moss or hay over the holes, and if sand is used for packing, spread on an inch or two in depth and over this a layer of the cuttings, then another thin layer of sand, and so on, until the box is full. When moss is used as a packing material, proceed in the same way as with sand. If a dry place can be secured in the open ground, the box containing the cuttings may be buried there and then covered with sufficient earth to prevent severe freezing during the winter. A little freezing will not injure the cuttings of hardy plants, though it may retard the production of buds and new roots; but a cool cellar is the best place in which to keep root-cuttings, because it will be a convenient place for examination from time to time, and their condition be ascertained as often as necessary. The greatest danger to be guarded against when cuttings are stored in a cellar is high temperature, which may force them into growth before the time arrives for planting out. The temperature of the cellar should remain only a few degrees above the freezing point, until the approach of warm weather in spring; then, if the callus and buds have not formed, increase the temperature as much as may be necessary to excite growth. It will be found in practice, that root-cuttings of certain kinds of plants produce a callus, buds, and even young shoots in a much lower temperature than others, hence the necessity of frequent examination. There is also a very marked difference in the readiness with which root-cuttings of the same genus produce buds, and very often, when two varieties of the same species are treated exactly alike, and placed side by side, one will produce buds and even shoots several inches in length, before the other has emitted a callus. I have often found this to be the case with root-cuttings of roses, and a little forcing with bottom heat was needed to make certain varieties produce buds. The intelligent propagator will readily see where a change in treatment is required, and bear in mind that it is only by close attention in such matters that success is to be attained.

In warm or hot climates root-cuttings may be planted out as soon as they are made, provided there is a certainty of sufficient rain falling to ensure a supply of moisture; but keeping them under artificial conditions will usually yield the most satisfactory results. In all cases the root-cuttings should be made when the plant has matured its growth of the season, whether it be of herbaceous or woody species, and the propagator should not forget that it takes time for buds to form from cell matter, and to force the roots when first divided is seldom or never advisable. In propagating tender or greenhouse plants, such as Bouvardias, Geraniums, Acacias and Coronillas, etc., by root-cuttings, the cuttings should be placed in the boxes and covered with sand, charcoal, or moss, and then set aside in a moderately cool place to callus and produce buds, before attempting to force them into growth.

Tender Roses, as well as hardy, may be rapidly propagated by root-cuttings, as is now generally practised with the Bouvardias, and at all seasons, provided the cuttings are taken at a time when the plants are in a semi-dormant condition. This may occur in the autumn, if the plants have been growing in the open ground, or in spring with those that have been forced in the house during winter.

Because certain kinds of plants may be propagated freely from ripe or green cuttings, it does not follow that they may also be propagated by root-cuttings. For instance, cuttings of the branches of the common Catalpa tree may be struck almost as freely as the Willow, but I have never known of a root-cutting having been made to produce a bud. I have kept the cuttings between layers of moss and in a greenhouse for twelve months, and at the end of that time they were perfectly sound and fresh, and while they had thrown out numerous rootlets, there were no indications of buds. J. C. Loudon, in all of his works where he has occasion to refer to this tree, says that it can be, or is, propagated by root-cuttings, and this statement has been repeated by various authors since his time, and I must confess to have been once misled and named the Catalpa among the ligneous plants which could be readily propagated by root-cuttings. The Paulownia tree, which is not distantly related to the Catalpa, is so readily propagated by root-cuttings, that nurserymen prefer this mode to any other. On the contrary, there are other woody plants, such as some of the Moss Roses, which almost defy the skill of the experienced propagator to force roots from the ripe wood, and yet cuttings of their roots grow very readily when treated in the same way as advised for those of the Raspberry and Blackberry. It is not to be supposed, however, that all of the known species of the Rose can be multiplied as readily from cuttings of their roots as the common Blackberry, but there are many that can be, and no doubt all may be propagated by root-cuttings by varying the treatment—some requiring more time and a higher temperature than others in order to secure the production of buds. I now refer to the true roots, not to what are called subterranean stems and branches, which are quite distinct productions from roots, as the former are similar in structure to the aerial stems and branches of the plants that produce them, having joints or nodes at which there is either a single bud or a concentration of a number of buds, as may be seen in such genera of plants as the *Calycanthus, Neviusia,* and such species as Ramanas Rose (*Rosa rugosa*), the Smooth Sumach (*Rhus glabra*), and various kinds of herbaceous plants, like the Canada Thistle, Toad-flax, some of the species of Phlox, *Spiræa, Pachysandra,* and others, all of which produce subterranean branches, in addition to their true roots. Furthermore, true roots always originate in the form of minute, tapering rootlets, enlarging with age, while the subterranean branch starts from a bud on the main stem, with nearly, or quite, as great a diameter as it will ever acquire, with a tip not pointed, but rather blunt, and often thicker than the part behind, and if it meets no obstacle in its lateral growth, the end will either come to the surface and produce an upright stem, and separate plant, as usually found about old clumps of the plants named, or they terminate in a tuber, as seen in the common Potato, Jerusalem Artichoke, Tuberous-rooted Scirpus (*Chufa*), etc. If the terminal bud on these subterranean branches is destroyed, the next below will produce a shoot, and this may be repeated until the entire number are forced into growth, as is well known to every farmer who has ever tried to kill out Canada Thistles.

The roots of the Raspberry and Blackberry, although often referred to as subterranean stems or branches, have really nothing in common with these productions, as they do not show any such system of nodes and internodes, with buds at their joints, as seen on their aerial canes and branches; but the roots of these plants produce adventitious buds from any and all parts of their surface alike, consequently in dividing them for making cuttings, we pay no attention to any surface indications of roots or buds. But in employing the true subterranean branches for cuttings, we divide them on the internode, preserving one, two or more joints with their dormant buds on each cutting. If one of these buds pushes and forms a sprout, it is sufficient, but sometimes all will grow, producing a greater number of plants, but less vigorous ones.

The after-treatment of cuttings made of subterranean branches should be very much the same as that given to

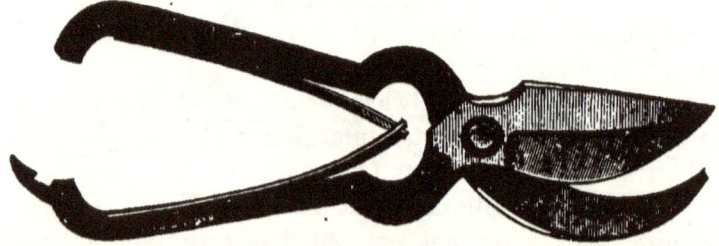

Fig. 69.—HAND-PRUNING SHEARS.

root-cuttings, although it will usually require less time for developing the dormant buds at their joints than to produce adventitious buds on true roots. In making such cuttings, the ordinary pruning shears (figure 69) may be employed, instead of a knife, for a very smooth surface to the severed parts is not so necessary as in making cuttings of the twigs and young shoots of ligneous and herbaceous plants.

To avoid repetition, I will say here that wherever in the following pages root-cuttings are mentioned as a mode of propagation, it will be understood that they are to be made and preserved as described in this chapter. Any variation in treatment which may be required will be noted as it may occur.

CHAPTER XVI.

PROPAGATION BY BUDDING.

Budding is the art of taking a bud, with a small portion of the bark adjoining, from one plant and inserting it in another, or in some other part of the same plant from which it was taken. The physiological principles which govern the operation are that there must exist an affinity between the plant from which the bud is taken and the one upon which it is to be placed, and the nearer the relationship the more readily will it unite and perfect the union. While the science of botany assists us in determining the relationships between plants, it is only by practical experience that the affinity between the various species can be determined. Two species of trees of the same genus may appear to be botanically very closely allied, and yet no permanent union can be effected between the wood of the two, and in such cases practical experience must necessarily become our only guide.

In budding it is very important that the bark of the stock should part readily from the wood; and to secure free and easy separation it is necessary that the operation should be performed when the flow of sap is abundant, because if the bud is inserted at this time it immediately comes in contact with that nourishment which it requires for its support. The sap which has been assimilated by the leaves descends mainly through the inner bark and on the external surface of the wood of the stock; it therefore comes in direct contact with the inside of the bark adhering to the bud, and is transmitted through the exposed cells to the bud itself, which thus becomes attached to the plant upon which it is placed; or, in other words, a union is formed between the two.

For budding are necessary a small knife for preparing the buds for insertion and opening the bark of the stock to admit them, and a quantity of some material to tie around the stock so as to hold the bud in place. Budding knives are made after various patterns; one that is commonly used has an ivory or bone handle, made very thin at the end, that is used to peel the bark from the stock where the bud is to be inserted, figure 70. Another form of budding knife is made with a horn handle, with a small tapering piece of ivory fastened in the end. These knives, of various sizes and shapes, can be had at the seed-stores, but another and quite a different form of budding knife is shown in figure

Fig. 70.—COMMONLY USED BUDDING KNIFE.

71, and known as the "Yankee Budding Knife." It is a small pocket knife with a thin blade, round at the end. The cutting portion extends about one-third around the end of the blade and about two-thirds of its length, leaving the lower part dull. Although this form of budding knife has been in use in some of our older nurseries for the past fifty years or more, still it does not appear to have been manufactured for the general trade, and only on special orders from nurserymen. I have

Fig. 71.—YANKEE BUDDING KNIFE.

used this knife for the past thirty years, and prefer it to those with a bone or ivory spatula for lifting the bark; for in using the Yankee Budding Knife there is no time wasted in reversing it, as is necessary with those of the form above mentioned. The rounded end of the blade is used for lifting the bark, and it is far more convenient than any form of knife that must be reversed in the hand every time a bud is inserted. This Yankee Budding Knife is an implement especially adapted to rapid work in the nursery. It may be said, however, that it is immaterial what form of knife is employed, provided it has a keen edge and is dexterously used. The material most commonly used for tying in the bud is called Bass, and may be procured at almost any seed-store, or it can be obtained in the form of Bass mats; but when it cannot be readily had in either of these forms, and Basswood trees are at hand, their inner bark may be stripped from them in the spring, as soon as it parts freely from the wood. By immersing this bark in water from two to four weeks (varying according to the temperature of the water), the bark will part with its mucilaginous matter, after which it may be divided into thin layers resembling fine silk, being very soft and pliable. Another good tying material is known in the trade under the name of Raffia or Roffia, and of late years it has been extensively imported for this purpose, and is now kept on sale at most seed-stores. It is the cuticle of the leaves of a large Palm, the *Rhaphia Ruffia*, indigenous to Madagascar and Mauritius. Raffia is somewhat softer and more pliable than the ordinary Bass bark but does not hold its form as well, being inclined to roll up instead of remaining flat when handled in tying. Bass, and similar materials, should be immersed in water for an

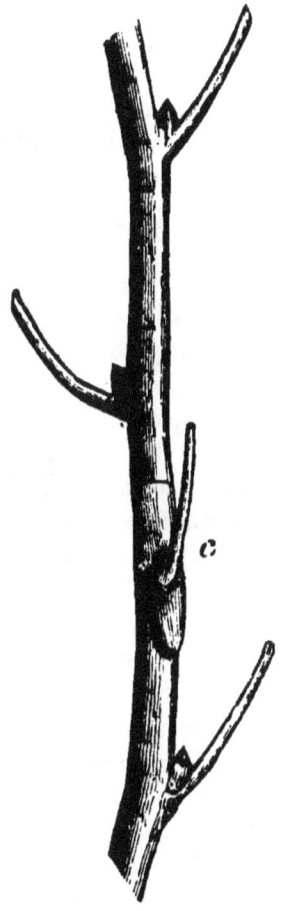

Fig. 72.—STICK OF BUDS.

hour or two before wanted for use, in order to increase their pliableness and softness. The inner bark of other kinds of trees besides the Basswood is sometimes used for this purpose, also woolen yarn, strips of cotton cloth, cotton wicking, etc., but Bass is more extensively used than any other material. In selecting buds the young shoots of the present season's growth are usually preferred, and these should be taken from the most healthy and thriving branches. The leaves should be immediately removed, leaving a portion of the leaf-stalk attached, as shown in figure 72. If the leaves have fallen from the twig, it is usually thought to be too ripe for use, but in some instances such buds may be used with success. If there are any soft, immature buds on the upper part of the shoot, or any undeveloped ones at the base, they should be rejected. But success in budding depends very largely upon the condition of the stock at the time the operation is performed. Unless the sap is flowing, and in sufficient abundance to allow the bark to part, or peel, freely and easily from the wood underneath, the bud is certain to fail. If the buds should happen to be a little over-ripe, or wholly dormant, if placed in direct contact with the living tissues and thin juices of the stock, they will absorb moisture and nutriment, and they are far more likely to unite and live than under opposite conditions.

To become an expert in budding, the following rules may be observed: Take the branch in the left hand with the small end pointing under the left arm; insert the knife blade half an inch below the bud, cutting through the bark and a little into the wood; pass the knife under the bud and bring it out above it, taking off the bud with the bark, and a thin slice of wood attached, as at *c*, figure 72. Then (if using the Yankee Budding Knife) let the forefinger clasp the lower part of the blade, make the horizontal incision in the stock first and from this an incision downward about an inch long, being careful not to cut too deep; lift up the edge of the bark by passing the back of the blade (without removing it) up to the horizontal incision. Lift the bark on the other side in the same manner, the two incisions making a wound on the stock resembling the letter T, as shown in figure 73. If other forms of budding knives are used the thin end of the ivory handle is thrust under the bark, raising it sufficiently to admit the bud. The budder holds the bud between the thumb and forefinger of his left hand while making the incision in the stock; and as the knife leaves it, he places the lower point of the bark attached to the bud under the bark of the stock before this falls back into place, and thrusts it down into position. If the upper end of the bark attached to the bud does not pass completely under the bark of the stock, it must be cut across so as to allow that which remains with the bud to fall into place and rest firmly on the wood of the stock, as shown in figure 74. When the bud is fitted to the stock, wind the Bass, or other material used, around the stock, both above and below, covering the entire incision, leaving only the bud and part of leaf-stalk uncovered. The ligatures should be removed or loosened as soon as the bud has become firmly united with the stock, which will usually be in ten or fifteen days, if at all. The horizontal incision is sometimes made below the perpendicular one. This allows more of the downward flow of sap to reach the bud than when cut across above it, as no cells are divided above the bud; but as it often proves detrimental, and is

Fig. 73.
INCISION READY FOR BUD.

Fig. 74.
BUD IN POSITION.

not so convenient, this mode is rarely practised, except upon plants in which some peculiar condition of the sap at the time of budding seems to require it.

When the bud is taken from the shoot, as represented in figure 72, *c*, there is a small piece of wood remaining under the eye, which in budding some kinds of plants it may be desirable to remove, although it is almost an universal practice in this country to let the wood remain, and doubtless in a majority of cases, and with most kinds of plants, it is best to do so; but there are many European propagators of plants who insist that a more permanent union can be secured by its removal. Certain French nurserymen claim that the removal of the wood is quite important in using the Quince as a stock for the Pear, preventing overgrowing or "knotting" at the point of union between stock and bud, but I do not think our own nurserymen have found in their experience that the removal of the wood in the ordinary "shield budding" is an advantage. It may be said, however, in its favor, that when buds are to be taken from large stock branches, like those produced by some varieties of the Pear and certain species of the Magnolia, that by removing the wood we secure a concave shield to set upon the convex surface of the stock, thereby making a better mechanical joining of the two than could be made otherwise. Various devices have been employed for removing the wood from buds, in addition to the more common one of lifting it out with the point of the budding knife, raising the upper end first, and peeling it downward to avoid breaking out the center or heart of the bud too deep, as is likely to occur if the lower end of the wood is lifted first and then pulled out from this direction.

But if the wood is to be removed from any considerable number of buds, branches should be used from which the bark will readily peel without tearing or breaking the fibres, and the buds removed as follows: Hold the

PROPAGATION BY BUDDING. 193

branch or shoot containing the buds in the left hand, and with the smaller end towards you; insert the knife blade about one inch below the bud; cut a little deeper than you would if the wood was to be left in; pass the knife above the bud about one inch, then cut across through the bark only, about half an inch above the bud (see figure 75), then with the finger and thumb lift up the bark, at the same time press gently forward, and the bark and bud (*a*) will come off, leaving the wood beneath (*b*) adhering to the branch. Examine the bud after it has been removed to ascertain whether the *chit*, as it is called, has been broken off even with the inside surface of the bark or within the bud, leaving a cavity; if the latter, there is danger that while the bark around the bud will unite with the stock, the bud itself may fail to grow unless the flow of assimilated sap on the stock is sufficiently abundant to fill the cavity with cambium soon after the bud is inserted. But the particular manner in which buds are taken from the twig, or inserted in the stock, will make but little difference, provided the buds are fresh and the operation is carefully performed at the proper time. One operator will insist that the best way to make the horizontal incision in the

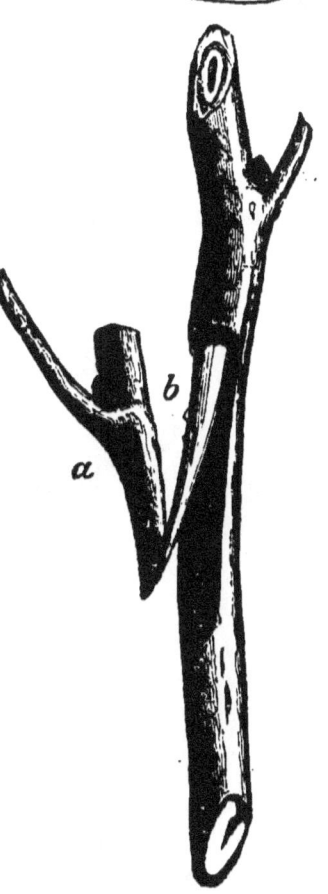

Fig. 75.—REMOVING THE BUD.

stock is by giving the edge of the knife a sloping inclination downward, as shown in figure 76. This form has its advantages, as well as disadvantages. The point of the bark to which the bud is attached is more readily inserted under the bark of the stock than when the incision is made square, but it also permits the water which falls upon the stock to enter the wound, often to the injury of the inserted bud. This method is given in "Warder's American Pomology," and is there called "Mr. Jackson's method."

The sloping incision would naturally suggest itself to any one who desired to insert buds very rapidly, because

Fig. 76.—BUDDING SLOPING OUT.

if the stocks were in a proper condition the buds could be thrust home without resorting to the knife for the purpose of lifting the bark; although it must be admitted that, as a general rule, to lift the bark with the knife is far better than to do it with the bud. The bark and wood to which the bud is attached are not usually firm enough to resist, without injury, the pressure required to cleave the bark from the stock.

The time for budding most kinds of hardy plants is usually during their growth in summer, varying the season to suit the different species, varieties, localities and soils. The experienced propagator, who has become acquainted with the habits of various species, usually delays the operation until the stock has passed its season of most rapid growth, for he has learned that if the bud is inserted too early in the season, the stock may overgrow and smother it. Then, again, buds which are set too early will often push into growth, and not having sufficient time to mature their wood before winter, are killed by cold. Pinching off the ends of the new shoots, when a few inches long, will usually check the growth and hasten maturity; but buds which make no growth the first season are in the best condition to resist the cold of winter. Removing the ligature with which the bud is tied, as soon as a union is formed between bud and stock, will usually prevent premature or fall growth.

Although summer is the best season for budding most kinds of plants, still, the operation may often be successfully performed in the spring, when vegetation is just starting into growth, or when the sap has begun to flow freely. The branches from which the buds are to be taken are usually cut from the parent plant early in winter and packed away where the buds will remain dormant and yet keep alive and uninjured until wanted for use. Budding in spring is performed precisely as in the summer, except that there should be no attempt at removing the wood. Buds may also be removed from one tree and inserted into another in the spring, if both stock and cion are in the same condition, but the operation is not generally as successful as summer budding.

With trees and shrubs which have a very thick bark, such as Hickory and Magnolia, or even those with a thinner bark, like the Chestnut, Cherry, etc., the annular or ring budding may be performed in the spring after growth has commenced, and in some instances it may be found convenient for saving some choice species or varieties, but it is too slow a mode for general use. It differs from other methods in several particulars, but the main one is that the bud is not inserted under the bark of the stock but fitted to it. A ring of bark passing nearly or quite around the stem upon which there is a bud is taken from the branch, and a similar ring is cut from the stock, and the bud and bark are fitted into this and then carefully tied in place. The branch from which the bud is taken, and the stock to which it is affixed, should be of nearly the same size, although a piece of bark may be taken off from the bud, or the same added to it, for the purpose of making a close joint. Figure 77 shows a stock with a ring of bark removed (*b*), and another (*a*) with bud ready to clasp around it. This method of budding or grafting, for it may be considered under either head, is termed "flute grafting" by European horticulturists, and it is more generally performed in spring than later in the season. When only a small section of bark is removed with the bud, and this fitted to the stock by removing a similar section, it is called "veneer shield budding," as it is intermediate between the ordinary budding and the annular or flute grafting.

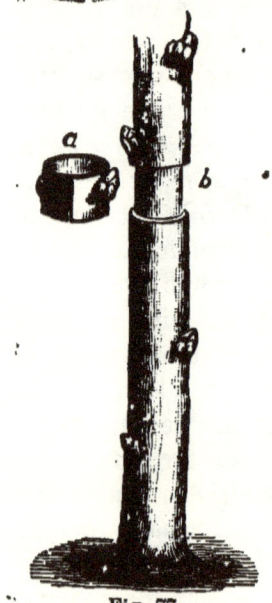

Fig. 77.
FLUTE BUDDING.

Like other methods of propagating plants, budding may be performed in various ways. The exact size or form of the bud or stock is not so very essential; the main thing to be observed is, their condition at the time the operation is performed. It should be borne in mind that new plant-cells must be formed before a permanent union can take place; consequently, both bud and stock should be in a condition to produce cells soon after the junction of these is artificially made. As the delicate tissues of growing wood and bark are very sensitive to exposure, it is apparent that in performing the operation of budding, in any of its forms, it should be done as rapidly as possible compatible with completeness of execution. The young wood containing the bud is, however, not so perishable as to prevent the preserving of them for several days in good condition, and it has recently become quite a general practice among nurserymen to supply their customers at a distance from the nursery with buds of various kinds during the season. When the sticks of buds, as they are termed, are cut from the tree, the leaves should be removed, leaving about an inch of the petiole, or leaf-stalk, attached to the branch for convenience in handling the bud at the time it is inserted in the stock. The sticks of buds should then be wrapped in damp moss that will keep them in good condition for several days, but the less interval that elapses between the time they are cut and that when they are used, the better. In the ordinary and more common modes of budding, during the months of July, August and September, in temperate climates, the bud is not expected to push into growth until the following spring, it remaining in a dormant state through the winter months. Whether the budding should be performed during the first or last month named, will depend upon the condition of the stocks, but those kinds that complete their growth earliest in the season, should be budded first. If the stocks are likely to cease growing before the buds are ready for use, the maturing of the latter may be hastened by pinching off the ends of the twigs, thereby throwing all the sap into that part which remains. The buds should be plump and of good form and substance, a condition which may be readily understood by an examination of shoots in different stages of growth.

If the stocks are very vigorous and are budded early, there will be danger that the ligatures may cut into the bark, as the stock increases in diameter. It would be safest to leave the ligatures on the stock until the following spring, provided the budding could always be done at just the right moment to insure a firm union, and no more, when frosts come to check growth, but this so seldom occurs, that in a majority of instances the ties will need to be loosened within ten or fifteen days after the buds are inserted. It is not always advisable to remove the ligatures, but merely to loosen them sufficiently to admit of an expansion of the stock, and prevent strangulation and the forcing of the bud to make a premature growth. If the buds should be forced into growth, there is nothing to be done but to head back the stocks and let the buds grow as rapidly as possible until checked by cold weather. Some of the lower buds on these premature shoots may escape destruction the ensuing winter; if so, they should be headed back to a strong, well-developed bud, and this may produce a strong and vigorous shoot.

Under ordinary circumstances, budded stocks should not be headed back until the following spring, and then severed at a point four or five inches above the inserted bud. This stump serves as a support to which the new growth may be tied, to prevent it from being broken out by the wind. All sprouts which may push out from the stock, either below or above the bud, must be rubbed off from time to time, as they appear. Later in summer, when the new stem has become somewhat mature, the stump of the stock may be cut away with a sloping cut, at the lowest point opposite the bud. With weak, slender-growing trees, the young plant may require staking for the first year or two, in order to secure a straight stem.

The rapidity with which budding can be done by an expert is astonishing to a novice or an amateur who will spend several minutes in properly placing a single bud.

I have known men who considered 200 an hour an easy task, and there are a few who will set many more than that, under favorable circumstances. Practice makes expedition as well as perfection.

CHAPTER XVII.

PROPAGATION BY GRAFTING.

Grafting is governed by the same physiological principles as budding. There must exist an affinity between the stock and cion; if not, a permanent union is impossible.

With some of the modes in use, the operation is very similar to that of budding, but with this important difference, that in grafting a larger section of the plant to be propagated is used than in budding; besides, it can be performed upon a great variety of plants while they are dormant.

The art of grafting is one of the most ancient methods known of multiplying individual species and varieties of plants; still, there is at this late day, scarcely one person in a hundred, among those who cultivate fruits who sufficiently understands the process to put it into successful practice. The same may be said of all the most common methods of propagating plants, oft-repeated, but seldom learned by any considerable number of persons of any one generation.

Through the ingenuity of horticulturists a great number of different modes of grafting have been invented, and practised to a somewhat limited extent, but of the majority it is safe to say that they are of no practical value, merely showing in how many different ways the cells of plants may be forced to unite. As it would be only a waste of time and valuable space to enumerate the many different modes of grafting, only those possessing the most merit, and such as have been approved by skilful propagators, will be described in the following pages.

There has also been a large number of grafting machines invented, several of them patented in this country and Europe, but practical propagators seldom employ them in their work. The French gardeners, and especially the vineyardists of France, have been quite prolific in inventions of this kind, and since the advent of the Phylloxera in the vineyards of Europe, grafting machines have been in greater demand, as they are thought to be of some advantage in grafting the Grape. As the species of Grape indigenous to North America are supposed to more successfully resist the attacks of the Phylloxera than the European varieties, the former are now extensively employed as stocks upon which to graft the latter. Among the various European implements recommended in grafting vines, the following are perhaps the best known kinds: M. Petit's Cleft Grafter; Leyder's Grafting Implement; Bordguer's Grafting Tool; Trabuc's Grafter; Sabatier's Implement for grafting in-doors, and another for outdoor work; Vincent's Grafter, and Pelanquier's Grafter. Some of these implements are of very complicated construction, and not only cut off and prepare the stock for the reception of the cion, but tie it in when inserted.

An American implement for grafting the Grape, consists of two saw-blades placed side by side, for cutting the cleft in large stocks, and an accompanying implement, consisting of two knife-blades set in a lever, for cutting the cions of the proper thickness to fill the sawcleft in the stock. This implement, although at one time highly recommended and somewhat extensively employed in grafting old and large vines, appears to have gone out of use, like others of its kind.

The ordinary implements used in grafting, are: a small

saw, for cutting off the heads of large stocks or branches of trees; a good, strong knife, with a thick back, to make clefts in the stock; a small knife with which to prepare the cion; a wedge, or grafting chisel, and a small mallet. The above-named implements are often made of peculiar patterns, to suit the fancy of the operator, but the main thing is, to have the work well done. Other kinds of implements are used in performing particular modes of grafting, which I shall have occasion to mention as the different processes are described. In addition to these, bass or raffia strings, such as are used in budding, for tying in the grafts, and grafting-wax, to cover the wounds and protect them from the air and water, are necessary.

GRAFTING-WAX.—There are many kinds of grafting-wax, as well as other compositions for the same purpose. A composition, made of clay, fresh cow manure, and fine straw or grass, was the principal material used in grafting until the present century, and it is still used occasionally, and with such good results that it should not be entirely ignored or overlooked. It may be prepared as follows: Take a quantity of good, strong clay, and a small quantity of fresh cow manure; add sufficient water to make it the consistence of thick paste; add a little fine-cut grass or straw; if a little salt—say about one pint to the bushel—is added, it will assist it in retaining moisture, when applied to the stock and cion. This composition should be made several weeks before it is wanted for use, and be thoroughly worked over as often as once a week until used, for the more it is manipulated the better. This composition is but little used at the present day, but for some kinds of coarse grafting on large, open-grained wood, it will retain moisture longer and protect the cion better than the more modern grafting-wax. There are many different kinds of grafting compositions recommended by the authors of works on gardening, which shows that the exact proportions of materials, or, in fact, the materials themselves, if of like nature, are not very essential to success. For grafting in the open air, the following compound is probably more generally used in this country than any other : Common rosin, four parts ; beeswax, two parts ; tallow, one part—melted together ; and after it is cool, it is applied by hand, or, when in a liquid state (being melted), it may be applied with a brush or spread thinly upon tough paper or muslin, and the latter cut up into strips of convenient size for use. If the wax is to be used in cool weather, then add a little more tallow. Linseed oil is sometimes used in place of tallow in the following proportions : Rosin, six pounds ; beeswax, two pounds ; linseed oil, one pint. From my own experience, I consider tallow much preferable to oil, and I would warn the novice against using indiscriminately the different kinds of oils often sold under the name of linseed. In Europe, Burgundy pitch is more generally used in making grafting-wax than in this country. Some of the French authors recommend the following : Melt together two pounds twelve ounces of rosin, and one pound and eleven ounces of Burgundy pitch. At the same time melt nine ounces of tallow ; pour the latter into the former, while both are hot, and stir the mixture thoroughly. Then add eighteen ounces of red ochre, dropping it in gradually and stirring the mixture at the same time. After the composition has cooled sufficiently, work it well with the hands. If this wax is to be used out of doors in cool weather, it should be carried in a vessel like an ordinary glue-pot, in order to keep it in a semi-fluid condition. All the above kinds of wax may be spread upon cloth or tough paper with a brush, when warm, and the paper or cloth cut up when the wax is cool. In what is called splice or whip-grafting, these waxed strips will be found very convenient for wrapping about the parts united.

The French Mastic, so long known as "Lefort's Liquid Grafting-Wax," is made by melting one pound of common rosin over a gentle fire, adding one ounce of beef tallow, the latter to be well stirred in. Take it from the fire, let it cool down a little, and then mix in eight ounces of alcohol. The alcohol will cool down the mixture so rapidly that it may be necessary to place it over the fire again. The utmost care must be exercised to prevent the alcohol from taking fire. The composition should be kept in tin boxes or glass jars until wanted for use. This mastic is highly recommended by the nurserymen of France, but as its composition and mode of making were until quite recently a kept secret, it has been used only in very limited quantities in this country, owing to the cost of importing it.

In all the different modes of grafting great care should be observed in having the external surface of the wood of the stock and cion exactly even—no matter whether the inner surface of the bark is even or not. This allows the new cells which form between the bark and wood, of both stock and cion, to unite and form channels through which the sap can readily pass. The sap ascends through the wood of the stock into that of the cion (graft), causing the leaves to expand, which, in their turn, assimilate it preparatory to its descent, as stated in a previous chapter.

The time for grafting most kinds of woody plants in the open air or nursery is in the spring, just before, or soon after, the sap begins to flow most rapidly, varying the time according to the nature of the different species to be operated upon, for experience has demonstrated that some kinds should be grafted much earlier than others, without regard to any apparent movement of the sap. These variations in the time of grafting, as well as in the condition of the stock, will be referred to more fully in a succeeding chapter.

In cold climates, the young slender branches of even the most hardy deciduous trees are often injured by the severe frosts of winter; therefore, when such twigs or branches are wanted for cions, it is best to take them from the parent stock in autumn, soon after the leaves have fallen, and preserve them in earth, saw-dust, charcoal, sand, moss, or some similar material, where they will be cool—not frozen—and just sufficiently moist to prevent shriveling. Cions of the ripe wood of some kinds of trees may be taken from the parent plant in the spring, at the time they are wanted for use; but their vitality is often weakened by the severity of the weather, and the delicate tissues injured to such an extent that they will not form what is called "granulations"—although it is precisely the same as the callus on cuttings—which fill up any small interstices that may exist between the stock and cion, allowing of the transmission of sap from one to the other. Furthermore, in grafting plants while in a semi-dormant state, it is well to secure as great a difference as practicable in the density of the fluids of the stock and cion, in order to ensure the endosmose and exosmose movement of the sap, as explained in Chapter I., and to secure this condition we have only to keep the branches selected for cions in a dormant state until the sap of the stock has begun to flow in spring. While it cannot be said to be positively necessary in every instance that there should be any considerable difference in the density of the fluids of the stock and cion, to insure success in grafting, still, with some kinds of trees, a difference in condition is rather to be sought than otherwise.

CLEFT GRAFTING.—This method is the original or most primitive of all the different modes of grafting trees, and it is principally used upon large stocks or on the branches of old trees. It is rather a bungling, unscientific method of grafting, exposing an unnecessary amount of surface to be healed over by a new growth, and the scars made in

PROPAGATION BY GRAFTING. 205

the operation are seldom or never entirely obliterated. It is, however, extensively employed in re-grafting old orchards, and in utilizing large stocks, in order to obtain bearing trees in less time than if smaller and younger ones were used.

The stock is first cut off at a point where it is desirable to insert the cion; it is then split with a large knife or thin chisel—being careful to divide the bark at the same

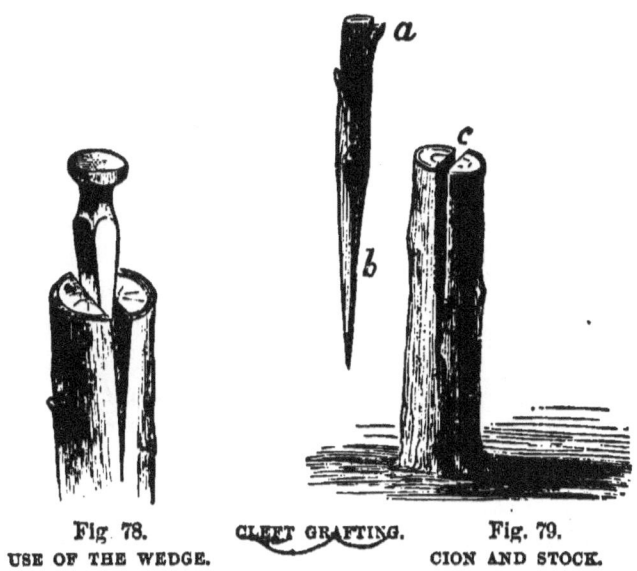

Fig. 78.
USE OF THE WEDGE.
CLEFT GRAFTING.
Fig. 79.
CION AND STOCK.

time, leaving the edges smooth. When the knife is withdrawn, an iron, or hard-wood, wedge is inserted in the center or at the side of the stock, as shown in figure 78; the cion, *a*, figure 79, is then cut in the form of a wedge, *b*, and fitted into the cleft, *c*; the wedge is then withdrawn, and the elasticity of the stock will hold it in its place. Grafting-wax is then applied, entirely covering the wound. When the stock is an inch or more in diameter, two cions may be inserted, one on each side, the operator being careful to place the external surface of the wound—not bark—of both cion and stock exactly even;

at least they should meet at some one point; and to make sure of this, some grafters set the cions slightly inclining inward, as shown in figure 80—*a*, the upper part of the cion; *b*, the lower end. The cion may be two or three inches long, containing one or more buds. The bark on the cion will usually be thinner than that on the stock; but this is of no consequence, provided the edges of the wound are even, bringing the bark of the cion and stock in direct contact. In stocks of less size, they may be cut off with an upward slope and the cion inserted on the upper or lower side; some grafters prefer one and some the other, but I have often set cions in both positions without discovering that either had any advantage.

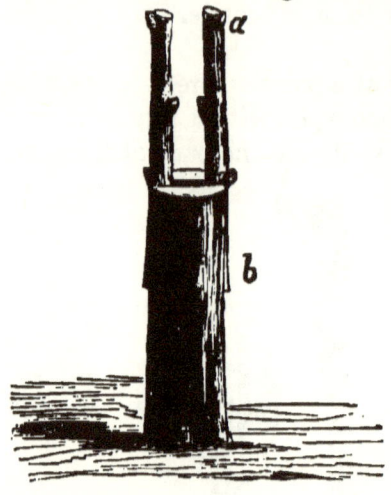

Fig. 80.
CLEFT GRAFT WITH TWO CIONS.

CROWN GRAFTING.—This is but a mere modification of the cleft graft, but instead of splitting the stock to receive the cion, the latter is sloped off thinly on one side, the bark divided from the top of the stock downward for an inch or more, and then lifted slightly, as in budding; the cion is then inserted under the bark of the stock and held in place by strips of waxed cloth. This form of grafting cannot be done until the bark of the stock will peel readily; consequently, it is usually performed later in the season than the ordinary cleft grafting. Another form of crown grafting is shown in figure 81. The cion is cut about half-way through, as shown at *B*, and the wood removed, leaving a square shoulder at top, and opposite to a well-developed bud.

From the stock, *d, d, d, d,* the bark is removed to admit the cion; one to four cions, as shown, may be fitted to a stock, and then all are held in place by a ligature of waxed cloth, and the top of the stock also covered with wax. This mode of grafting is usually considered best adapted to large stocks, such as are not suitable for the ordinary cleft grafting, but it may be used for stocks of quite small size. The exposed wood of the upper end of the cion should also be covered with wax, to prevent evaporation of moisture therefrom, and with some kinds of trees,

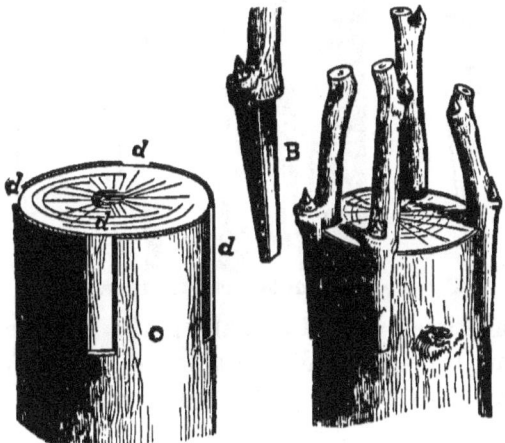

Fig. 81.—CROWN GRAFTING.

such as the Hickories, Walnuts and Chestnuts, it is well to cover or enclose the entire head of the stock and cions with a hood made of oiled paper or thin muslin, until the buds on the cions push into growth. This shading and protection against drying winds, often secures the growth of the cions when, if left exposed, they would fail.

TRIANGULAR CROWN GRAFTING.—This is only another form of the preceding mode, and one that should have long ago taken the place of the more clumsy method of cleft grafting. In this, the stock is not split, but instead, a triangular incision is made in the side of the stock, as shown in figure 82, *r,* and the cion cut in the same form and fitted into the cleft, as seen in the right-hand figure. An implement, called a cleft-cutter, figure

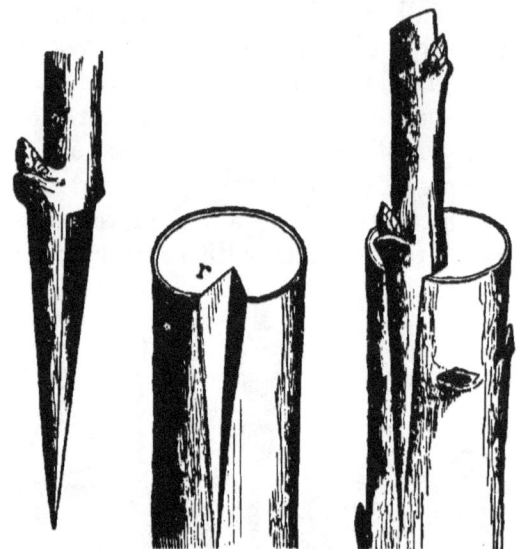

Fig. 82.—TRIANGULAR CROWN GRAFTING.

83, is sometimes used for making the incision in the stock; but it is not indispensable, as the cleft can be made almost as readily with a good, sharp knife. The

Fig. 83.

cions are, as is usual in such methods of grafting, held in position by ligatures of waxed cloth.

SIDE CROWN GRAFTING.—This mode of grafting is employed principally on large stocks and at the collar or crown above the main roots, and with species that cannot be readily divided or split to receive the cion, as in the ordinary cleft grafting. Sometimes the wood at this point is gnarled and so cross-grained that a smooth cleft cannot be made with a knife, and with such the side

PROPAGATION BY GRAFTING. 209

crown graft may be employed. The stock is cut off level with, or a little below the surface of the ground, as in figure 84. The cion, *B*, is severed to about two-thirds of its diameter, and this portion removed, forming a shoulder at *C;* the remaining part is pared smooth and thin at the lower end. The stock having been cut off at *D*, and the bark, *E*, removed with a thin slice of wood,

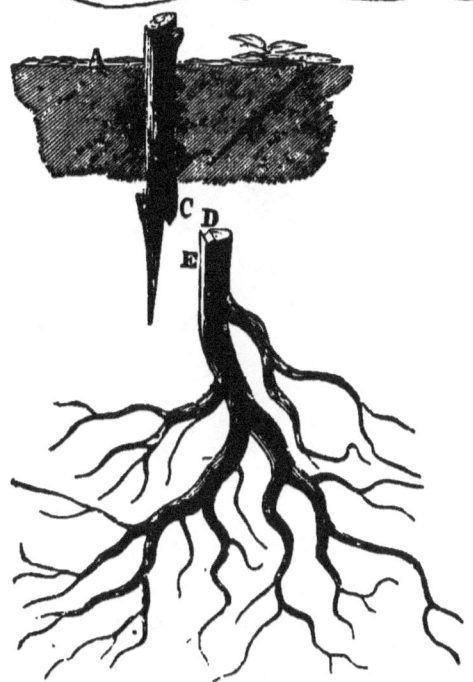

Fig. 84.—SIDE CROWN GRAFTING.

to correspond with the lip of the cion, which is then fitted to it, the shoulder of the cion resting upon the top of the stock, or both may be notched, as shown in figure 84. The cion and stock are then tied with bass or waxed cloth and the soil drawn back over the head of the stock covering the cion, except the terminal bud, *A*. It often occurs that cions of rare varieties of ornamental trees are obtained by persons who have no suitable stocks on which to use them, although large trees of closely allied species may be near at hand. Under these circumstances the large roots of old trees may be employed as stocks without removing them from the ground. In using such stocks, the soil should be removed from over the roots several feet from the main stem of the tree, varying the distance according to the age and probable size of the roots sought. The root is then cut off and the end brought to the surface, as shown in figure 85, and the cion inserted or affixed in any of the different modes of crown or cleft grafting. It must be apparent that a large root of this kind will supply a cion with ma-

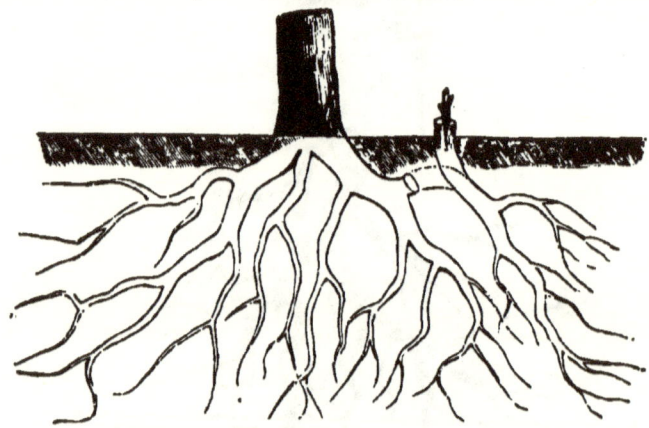

Fig. 85.—SIDE CROWN GRAFTING ON ROOTS.

terials for making a vigorous growth, which may be utilized as cions when a better class of stocks have been raised or otherwise procured.

SADDLE GRAFTING.—This is a neat but somewhat tedious mode of grafting, consequently seldom practised, except by amateurs in gardening who can spare the time necessary for such complicated operations. It is employed principally upon small stocks, or on the terminal shoots of young trees. The stock and cion should be nearly of the same size, although the stock may be a little

PROPAGATION BY GRAFTING.   211

larger without making any material difference in the result; the two sides of the stock are cut off in a sloping direction, forming a wedge, as shown in figure 86; the lower end of the cion is trimmed out on the inside so that it will fit upon the stock, as shown. Young Apple, Pear, and other fruit trees, when three or four feet high, are often top-grafted in this manner, although what is

Fig. 86.—SADDLE GRAFTING.   Fig. 87.—MODIFIED SADDLE GRAFTING.   Fig. 88.—KNIGHT'S SADDLE GRAFTING.

called a splice graft will answer equally well, and can be applied far more expeditiously.

Sometimes the saddle graft is so modified that it is intermediate between the cleft and the saddle, as shown in figure 87.

Another form of saddle grafting, introduced by Mr. Thomas A. Knight, of England, in 1811, is shown in figure 88. Of this Mr. Knight says that "it is never attempted until the usual season of grafting is passed, and the bark is readily detached from the alburnum. The head of the stock is taken off by a single stroke of the knife obliquely." The cion, which should not exceed in diameter half that of the stock, is then divided longitudinally, about two inches upward from its lower end, into two unequal divisions. The stronger division of the cion is then pared thin at its lower extremity, and introduced, as in crown grafting, between the bark and wood of the stock, and the more slender division is fitted to the stock upon the opposite side. The cion, consequently, stands astride the stock, to which it attaches itself upon each side, as in the more common mode of saddle grafting.

SPLICE AND TONGUE GRAFTING.—When the stock and cion are nearly of the same size, splice grafting—also called "whip grafting"—is the most convenient and certain method known. Seedling stocks are most generally used, and of various ages, from one to three or more years old, according to their kind and rapidity of growth. The stock is cut off with an upward slope, making the exposed wood perfectly smooth; a cion two to four inches long is cut off, with the same slope as the stock, and fitted to it, being careful to have the wood and bark on one side exactly even. The difficulty in practising this mode is in keeping the cion in position while applying the ligatures, and for this reason it has been almost entirely superseded by

TONGUE OR WHIP GRAFTING, which differs from the ordinary splice only in one point, viz.: a small cleft or split is made in both stock and cion, about midway on the slope, forming a tongue on both; these are then inserted one into the other, which will hold the cion in its place. Figure 89 shows the operation as completed—*c*, the stock; *b*, the cion; *a*, bud on cion—the union being formed by what is sometimes called a tongue splice. This mode of

PROPAGATION BY GRAFTING. 213

grafting is one of the most expeditious and certain of any in general use, not only where small seedling stocks and roots are employed, but also for top-grafting young trees in the nursery. It may also be used upon quite large stocks, provided they have been previously headed back and have thrown out thrifty branches upon which the cions can be conveniently spliced. In splice grafting, in the open air, waxed cloth should be used for ligatures, to prevent the breaking away of the cion before it has become firmly united; but when the grafting is done inside, or upon pieces of roots which will be planted out in spring, strong waxed paper may be used, or even bass, and no wax. Sometimes a ligature of bass is used, and then melted wax is applied with a brush, to cover the wounds on the joint between cion and stock. In some forms of root grafting it is desirable to place the cion so low down that it will eventually take root and become capable of supplying itself with sustenance through its own roots, instead of through those of the original stock. Under such conditions only very short pieces of roots are employed, as they are intended only to serve a temporary purpose, usually dying when the cion has produced roots for self-support.

Fig. 89.
TONGUE OR WHIP-GRAFTING.

All the other modes, such as the splice, cleft, crown, side and saddle grafts, may be used on roots as well as on the stems and branches of woody plants, and, as a rule, the larger and stronger the stock, the more vigorous will be the growth of the cion. But while rapid and vigorous growth is usually desirable, it should be kept in mind that no cion can utilize more nutriment sent forward by the stock than its leaves can assimilate; consequently, if the roots of a stock upon which a cion is set gather more materials than can be used, there must ensue a forced, unhealthy growth, or an entire inaction, in some part of the plant. When a tree or shrub is severely headed back, for grafting or other purposes, and thus deprived of its usual amount of foliage, it will often expend the greater part of its vitality in producing suckers or sprouts from the base of its stem. No root will remain dormant and healthy for any considerable time under circumstances which are naturally favorable for promoting growth, and, knowing this, the grafter should avoid cutting off all the branches of a large tree at one time, unless he can substitute a sufficient number of cions to fill their place, or at least enough to allow all the roots to act, even if it be but slowly. To avoid the too severe checking of root action, it is better to graft only a part of the branches of large trees one season, leaving the remainder until the next.

VENEER GRAFTING.—This mode of grafting is principally employed in propagating woody plants under glass, where both the temperature and hygrometric condition of the atmosphere can be readily controlled by the propagator. The usual time for performing the operation is in summer, and soon after the first and most vigorous growth of the season is completed, but before the wood and leaves are fully mature. As the leaves on both stock and cion are retained, they should not have entirely ceased to assimilate sap at the time of grafting, but still remain fresh and capable of performing all of their natural functions. The time for performing the operation must necessarily vary with different kinds, according to the difference in the natural habits of the various species, as some make their growth much earlier in the season than others, but the method of grafting is the same in all.

In this mode of grafting the stocks should be grown in pots for convenience in handling when performing the operation, as well as afterwards, for the union between stock and cion must be secured before the plants are removed to the open ground, or to outside frames. The usual practice is, to place the stocks in pots, from six to twelve months before they are wanted for use, and then plunge them in an open border surrounded with board frames, where water can be applied as often as necessary to ensure a vigorous growth of the stems and the formation of new roots. Success depends very much upon the condition and vigor of the stocks, and their preparation for use is of such importance that it should not be overlooked or neglected. Seedling stocks are principally used, and of various ages, according to the kind and natural growth, but those of from six to twenty-four inches high will usually be as large as necessary. If the seedlings have long tap-roots, these may be cut away, and even the lateral roots may be shortened, if necessary, to admit all into pots of convenient size. The tops may also be headed back at the time of placing the stocks in pots; in fact, more or less pruning will usually be necessary, in order to secure neat, trim-looking stocks. The new growth which they will make in the pots before they are wanted for use may require slight attention, in order to secure a smooth, clean surface on the stem at the point where the cion is to be placed.

When ready for grafting, the stocks are lifted from the border and carried indoors, and the cions cut from the parent plants as required, and kept in as fresh condition as possible, not being allowed to wilt or shrivel in the least. The stock should not be headed back or severely pruned at the time of grafting, although a side branch, or more than one, may be removed if necessary in preparing a place for the cion.

In affixing the cion the operator selects a smooth place on one side of the stock, then with a sharp knife he makes a light cross-cut through the bark and to a slight depth into the wood underneath, then inserts the blade from one and a half to two and a half inches above, cutting off a thin slice or veneer of the wood and bark down to the cross-cut in the stock. A similar slice is then cut from the cion, as shown in figure 90. The exposed alburnum of the cion is placed against that on the stock; and the whole wound firmly bound with a ligature of bass, as in budding or splice grafting. No wax is used, neither is there a tongue made on either cion or stock, but merely a clean, smooth wound, as shown in figure 90.

As soon as the cion has been inserted, the stock should be removed to the inside frames of the propagating house and gently watered overhead. The bottom of the frames should be either covered with sand or moss—the latter is preferable, as it holds moisture better and gives it off slowly, keeping the air within the frames well filled with vapor. If the frames are not deep enough to admit the grafted plants when set upright, the pots may be tilted over to one side, and a good depth of sand or moss in the bottom will aid greatly in keeping them in this position.

The house in which plants are veneer grafted in summer should be well shaded, either with lath screens or whitewash on the glass, and in very clear weather it will frequently be necessary to add extra shading to the inside frames, especially if filled with recently grafted plants of broad-leaved kinds, like the Maples, Magnolias and Dogwoods. In this climate, artificial heat will seldom be required; still, it is well to have the furnaces in order, as cold storms occasionally occur, and a little fire heat may be needed to allow of rather more ventilation than could otherwise be given with safety. During the first few days, or for the first week after the cions are set, the plants should be kept in a pretty close, warm and moist atmosphere, for the object at this time is to excite growth in the stock, or at least to accelerate the flow of sap, in order to produce rapid granulation of the wounds on both stock and cion, and thereby increase the chances of a union of the two.

Fig. 90.—VENEER GRAFTING.

Within a week after the cions are set, the failures, if any, may be seen, as the leaves will drop from the cions if there is no communication between them and the stocks. At the end of two weeks, or a little more, the cions will be united, if at all, and a very reliable indication of this is a new growth on the cions. The plants may then be removed from the house, if necessary, and the pots plunged in frames in the open ground, and given shade and plenty of moisture. But where the propagator has abundant house room, it is best to keep the plants inside for a month or more, and by proper ventilation somewhat harden the new growth before removing to the outside frames. If the stocks have a good ball of roots they may be slipped out of the pots when transplanted to the outside frames.

This mode of inside veneer grafting is not only one of the most expeditious and certain methods of propagating nearly all kinds of deciduous and evergreen trees and shrubs, but it is performed at a season when the expert grafter has the most leisure, if it can be said that such a time ever falls to his lot; and by having his stocks of various kinds growing in pots, he can graft them at his pleasure, in foul or fair weather, and at the same time be eminently successful in propagating many species of trees that are always uncertain, and seldom or never successfully grafted in the open air—at least not in cold climates.

Another advantage of grafting in summer is that in taking the cions, the propagator can select those which will best perpetuate any special characteristic—like the variegation of leaves in deciduous trees, which become obliterated later in the season. This is a very important matter, for, as is well known, the best marked of the variegated, laciniate and colored leaved trees, are inclined to revert to the original forms, and it is only by proper and timely selections of wood for propagation that these abnormal forms are fixed and perpetuated. By repeated selections of the best forms and most distinctly marked, these peculiar characteristics—which in a majority of instances give to the plants their special value—become intensified.

In the after treatment of the veneer-grafted plants, much will depend upon the climate where they are grown. If the winters are severe, they may need protection under glazed sashes, board shutters, or lath screens, with plenty of hay or some similar material filled in around the plants to prevent injury from alternate freezing and thawing or low temperature. The propagator must necessarily be the best judge of the amount of protection required, if any is needed.

The stock above the cion should not be removed until the following season, and in some of the very slow growing kinds it may be well to leave it intact until late in the summer or autumn.

BOTTLE GRAFTING.—While the mode of veneer grafting described is undoubtedly the best, and most readily and rapidly performed, others are sometimes employed; probably more to show how many different ways there are of obtaining similar or the same results, than for their practical utility. What is called "bottle grafting" is one of these variations from the more general method of veneer grafting. In this, the lower end of the cion, instead of being fitted to the stock, is placed in a bottle filled with water, as shown in figure 91. The upper part is fitted to the stock in the same way as in the former mode.

Another and rather more complicated form of bottle grafting is shown in figure 92. *A E*, the stock; *B*, the cion; *D*, ligature; *H H*, branches and leaves of the head of the cion; *F*, bottle filled with water, and *G*, stake to which the bottle is tied. This latter form may answer for amateurs who may wish to graft an Orange, Lemon, Oleander or similar specimen plants, but the professional gardener will seldom have occasion to resort to such complicated and time-consuming methods.

SIDE GRAFT WITH VERTICAL CLEFT.—There are propagators who prefer this method to the veneer graft, and it is usually performed under the same conditions. The cleft in the stock is made sloping downward and inward towards the center of the stock, cutting nearly or quite through the alburnum. The cion is made in the form of

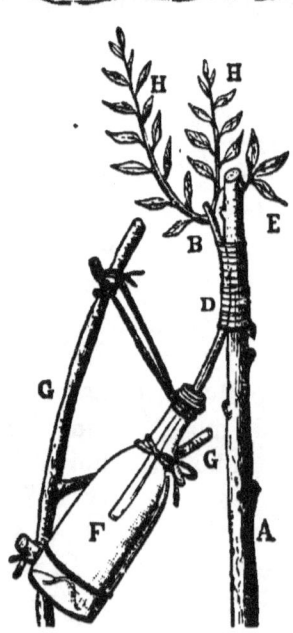

Fig. 91.
BOTTLE GRAFTING.

Fig. 92.
BOTTLE GRAFTING, MODIFIED.

a thin wedge and thrust into the cleft and held in place with the ligature of bass. This form of side graft is usually employed in what is called herbaceous grafting, as used in grafting Melons, Squashes, Dahlias, Potatoes, Tomatoes, etc. The cion, when grafting such succulent plants, is not pared down so thin as with woody plants, but left more in the form shown in figure 93—*A*, stock;

B, cleft in same; *D*, cion; *E*, leaf on cion; *F F*, leaves on stock. In grafting herbaceous plants and vines in the open air, heavy shading and frequent applications of water to the foliage are usually required to ensure success.

GRAFTING BY APPROACH.—This method is practised both in the open air and under glass, as all that is required is to have the stock and the plant that is to furnish the cion in sufficient close proximity to admit of the branches being brought together. A thin slice of bark and alburnum, two or more inches long, is removed from each, and the exposed wounds brought together and held firmly in place with a ligature of waxed cloth or bass; but if the latter is employed, it should be covered with wax if in the open air.

INARCHING.—This method differs from the last only in the manner of manipulation. To graft trees by inarching or approach, they must necessarily stand so near together that their stems or branches can be united without separation from the parent stock. Incisions are usually made similar to those employed in tongue grafting. The branches of different trees or of the same tree may be inarched, and in this manner hedges and live fences and screens may be formed with fruit or ornamental trees and shrubs. Inarching is sometimes employed in propagating rare species, instead of grafting in the ordinary methods; and after the union has been formed, the inarched branch is separated from the parent stem. In former times this method of propagating trees that were

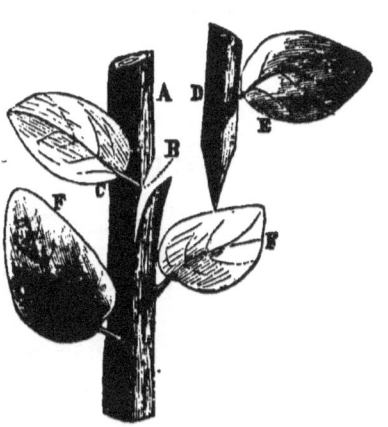

Fig. 93.—SIDE GRAFTING.

supposed to be difficult to graft in other ways, was very extensively employed by nurserymen. Seedling stocks were planted around a large or medium-sized tree, and near enough to allow the branches to conveniently reach the stock when ready for use, at which time they were inarched, afterwards severed, and the stocks taken up and removed to the nursery rows. The time for inarching trees in the open air is in spring, at the usual season for grafting, but if grown under glass, the operation may be performed whenever the plants are growing or are about to commence growth.

CHAPTER XVIII.

SELECTING STOCKS.

In the propagation of plants by budding and grafting, the selection of the proper kind of stock is quite as important as knowing how to utilize it afterwards. The most skilled propagator cannot produce the best nor even satisfactory results with poor stocks and cions. A feeble cion may revive and make a fair or even first-rate plant if set in a sound and vigorous stock, but if the stock is poor, the result is usually unsatisfactory, for in this case we build upon a feeble and unstable foundation.

It is not only desirable to secure stocks of closely allied species, but those that are young, vigorous, and well supplied with fibrous roots, for it is the small roots that first imbibe nutriment from the soil, as well as the first to emit additional fibres to assist in increasing the supply.

As a rule, seedling stocks are preferred to those raised from cuttings, although there are a few exceptions, and the "wilding," as it is termed, is usually the most hardy and least subject to disease. The so-called improved, domesticated and long-cultivated plant has also been longest exposed to diseases which often originate under forced and unnatural conditions.

In raising all kinds of tree and shrub stocks from seed, a moderately light, porous soil is preferable to one of an opposite character, as light soils tend to increase the number of fibrous roots. In heavy, compact soils, seedlings may produce longer vertical or tap-roots and taller stems, or we may say that they will grow larger in a given time than in lighter and naturally poorer soils, but they will usually produce a far less number of fibrous roots; consequently, less valuable for transplanting. The taller the stem and longer the tap-root, the greater the amount of each will have to be cut away when they are transplanted. The best stocks, therefore, for all purposes, are those that will lose the least when prepared for re-planting in nursery rows or for potting. The raiser should seek great diameter of stem and number of rootlets rather than length at the sacrifice of breadth. Thick seeding tends to lessen diameter and increase length, and while a greater number of plants may be produced on a given space by crowding, it is always at the expense of quality; consequently, height of stocks is not a safe guide for determining their vigor or value, either in seedling stocks or older trees.

There are certain kinds of trees, like the Oaks, Chestnut, Hickories, and English Walnut, which produce rather long, sturdy vertical roots the first season, and these poorly furnished with fibres, especially if raised on a firm, hard soil; but if the nuts are planted in light vegetable mold or sand, their roots will be mainly fibrous, with only a small central or tap-root, or none at all. Nurserymen who make a specialty of raising such stocks for grafting in pots, sow the nuts in shallow pots or boxes, and in nearly pure sand, applying liquid manure as needed, to insure a vigorous growth.

STOCKS FOR FRUIT TREES.

The stocks principally employed in propagating fruit trees of temperate and semi-tropical climates are as follows :

THE ALMOND.—Seedling Peach or Plum stocks are preferred. Seedlings of the hard-shelled Almond will answer equally as well as the Peach. The Plum makes the best stock for trees to be cultivated on moist or clay soils, and it is also less liable to the attacks of insects and diseases, but does not grow so large, and there is danger of the Almond overgrowing and dying through what may be termed strangulation, unless very large-growing varieties and species of the Plum are employed as stocks.

APPLE.—Seedling stocks are always preferred for general use for orchards, and those raised from ungrafted or wildings of the European species (*Pirus Malus*) are best. Stocks for the different varieties of the Siberian Crab Apples (*P. prunifolia*), may be of the same species, or seedlings of the wild American Crab Apple (*P. coronaria* and *P. angustifolia*), but those of the common European Apple are most extensively employed for all the cultivated species.

For dwarfing the Apple where low bushes or espalier training is desired, the Doucin and Paradise stocks are employed, these being low-growing varieties of the *Pirus Malus*. These stocks are raised by cuttings, or by banking up around the sprouts, which spring up around the base of large stems of plants which have been previously headed back for the purpose of producing them.

APRICOT.—Seedling Plum stocks, or those raised from cuttings, are usually employed in propagating the improved varieties of the Apricot. In mild climates both Apricot and Peach seedlings are sometimes used.

CHERRY (*Cerasus*).—For large, standard trees, seedling stocks of the wild Mazzard Cherry of Europe (*Cera-*

sus avium), are the best. In cold climates they should be worked low down near the crown. What are called the Morello, Duke and Kentish varieties of the Cherry are supposed to have originated from a wild species in Europe, known as *Cerasus caproniana*, and as a class they are considered more hardy than those that are descended from the *C. avium*. Seedlings of each group may be employed as stocks for their varieties, but the Mazzard is the largest and most free-growing tree, and for this reason is usually preferred as a stock for all. The Mahaleb or St. Lucie Cherry (*C. Mahaleb*), is a low-growing, slender-branched species, which is extensively employed in France as a stock for dwarfing the taller-growing varieties of other species. It has also been employed—but less extensively—for the same purpose in this country, but it should never be used for what are termed standard trees, or trees with long stems, for if trained high, the leading branches soon become diseased, die back, and if the tree does not perish altogether, it will assume the low dwarf form of the stock. Buds of the different varieties of the cultivated Cherry take very readily on Mahaleb stocks, and usually make a most vigorous growth for the first year or two. This apparent vigor of the young plant has often misled the inexperienced to believe that it would continue in after years if the tree is pruned up as a standard, but the chances are ten to one against securing such results.

Among the Cherries proper, or *Cerasus*, there are two very distinct groups of species which do not appear to have the least affinity, either in their flowers or wood; consequently, no hybridizing, so far as known, has ever occurred between the species of the two groups, nor have the plants of one been used as stocks for the other. These two groups are readily distinguished by their flowers; for in one they are produced in sessile umbels, as seen in the common garden Cherries from Europe, also in our Wild Red Cherry (*C. Pennsylvanicum*), and the Dwarf Cherry (*C. pumila*). The flowers in the other group are in long slender racemes, as in the American Wild Black Cherry (*C. serotina*), Choke Cherry (*C. Virginana*), Bird Cherry of the Rocky Mountain regions (*C. demissa*), and the small Bird Cherry of Europe (*C. Padus*).

The Chinese and Japan Cherries belong to the first section, and seedlings of the different species may be interchanged in their propagation whenever desirable.

CITRON, LIME, LEMON, ORANGE, SHADDOCK, ETC.— As all of these fruits are but different and closely allied species of one genus, their wood readily unites by either budding or grafting. But, as with other fruits, the largest and most vigorous species and varieties are preferred for stocks if large trees are desired, and the reverse for dwarfs; and as with the Apple and Pear, seeds from the Wild Orange produce better stocks than those from the improved on which to graft, and the same may be said of the Lemon and other species of the Citrus Family. The Shaddock (*Citrus decumana*), is a rather strong and large growing tree, and it will readily take buds from the Orange, Lemon and other species of the genus. For making dwarf trees of the Orange, the *Limonia trifoliata* is by some propagators considered a better stock than the Otaheite Orange, which has long been used for this purpose.

CORNEL TREE.—Seedlings of the common *Cornus Mas* are used principally as stocks for the different varieties of the Cornel, or edible-fruited Dogwoods. But the Cornels are so rarely cultivated in this country, that experiments are wanting for determining the value of the different species for stocks on which to grow the edible fruited varieties, but it is quite probable that *Cornus florida* would answer equally as well as or better than *C. Mas*.

CURRANT.—Although not a tree, the Currant is sometimes cultivated as standards and in the form of small trees. To secure this form they are grafted upon some strong growing species, like the Missouri Currant (*Ribes aureum*), which appears to answer the purpose as well as any that has thus far been tested.

DATE PLUM.—See Persimmon.

FIG.—The Fig is rarely propagated by grafting, as it grows very readily from cuttings; but weak and feeble varieties may be grafted upon the strong and vigorous.

GOOSEBERRY.—Like the Currant, this shrub is sometimes trained in the tree form, and the low growing varieties grafted upon the tall and most vigorous. Stocks raised either from seed or cuttings of the wild North American species are the best for this purpose. The Round-leaved Gooseberry of the Western States (*Ribes rotundifolium*), is one of the strongest and tallest growing of the indigenous species, consequently one of the best for stocks.

MEDLAR.—Strong growing species of the Hawthorn are preferred for stocks, but seedling Medlars, or even the Quince, may be employed for this purpose.

MULBERRY.—Seedlings of the common White Mulberry (*Morus alba*), are usually employed as stocks upon which to graft the Downing and other improved varieties.

NECTARINE.—As the Nectarine is only a smooth-skinned Peach, the same kinds of stocks and treatment are recommended for both. (See Peach.)

OLIVE.—In Southern Europe wild seedling Olives are often used as stocks upon which to graft the cultivated varieties as a means of hastening maturity. Grafting the Olive is not, however, a very general mode of propagation, as all the varieties are easily propagated by cuttings of the old as well as young wood.

PAPAW (*Asimina*).—Seedlings of the common Papaw (*A. triloba*), may be employed as stocks for the smaller growing species, or for the multiplication of distinct varieties; but of the latter, few or none have as yet been discovered worthy of a place among choice garden and orchard fruits.

Peach.—Seedling Peach stocks are principally employed in propagating the improved varieties. The best stocks are raised from what may be considered the inferior varieties, or, as usually termed, "wilding," that are on their own roots, not having been budded or grafted. Of late years, in this country, some care has been necessary in selecting seed in order to obtain it from trees that were free from a common disease of the Peach known as "yellows." Seedlings of one season's growth are preferred to older ones, and if the Peach stones are planted in moderately rich soil in spring, they will usually produce stocks strong enough to receive a bud by the ensuing August or September. If the bud "takes," the stocks may be headed back the following spring to within four inches of the inserted bud, and later in the season cut down close to the base of the young shoot produced from the bud. In planting the Peach stone or seed, they should be dropped about one foot apart in the row, or at a sufficient distance apart to admit of budding at the proper time.

Plum stocks are also employed for the Peach, especially in Europe, where the trees are trained to walls and kept low in the form of dwarfs, or cultivated in peach-houses under glass. The Plum is naturally a slower growing and a smaller tree than the Peach, hence its influence as a stock is to dwarf the growth. The Plum stock is no doubt preferable where the trees are to be trained low, or planted in stiff, cold, or clayey soils; but where the Peach thrives as an orchard tree, as it does in the light, warm soils of our best peach-growing districts, the natural stock is no doubt the best, although not quite so hardy or free from insect enemies as the Plum.

PEAR.—Seedlings of wild or inferior varieties are preferred to any other as a stock for the Pear. Most of the seed used in this country is imported from Europe, where it is saved from the pomace after expressing the juice of Pears for making perry. In raising seedlings in this country, great care is required in their cultivation to prevent blight on the leaves during the first season. Fresh, new, or virgin soil is best for a seed-bed, and it should be worked very deep in order to insure an equable amount of moisture during the entire summer. While seedlings of the Wild Pear of Europe have long been considered as the best stocks for standard trees, it is quite probable that the oriental species (*Pirus Sinensis*), and some of its varieties, found in both China and Japan, will eventually prove to be superior to the European species as a stock upon which to work the improved and long-cultivated varieties. These oriental species and varieties appear to possess a vigor surpassing that of any of those of European origin, as seen in their large, thick, leathery leaves, as well as in the great rapidity of growth and sturdy character of their young shoots. They certainly give promise of value as stocks upon which to work the less sturdy-growing varieties.

The Quince has long been used as a stock for the Pear, especially for producing low-growing or dwarf trees, which are desired for small gardens, or for training in some other form than as standard trees. While some varieties of the Pear succeed admirably when grown on Quince stocks, others soon fail for the want of a proper affinity between the wood of two species of trees which we force to unite temporarily, or otherwise, by budding or grafting.

By adopting a method known as "double working," we may use Quince stocks for varieties that do not succeed when worked directly upon it. In double working, varieties are selected that are known to unite readily and grow thriftily upon the Quince, and buds of these are inserted in the usual way, and near the ground. The next season, after the shoots from the buds have nearly completed their growth of the season, buds of the feeble-growing varieties, or those which appear to have very little affinity for the Quince stock, are inserted in the young growth of the Pear wood six to twelve inches above its junction with the Quince. The stock is then treated as it was the year previous, and the following spring the stock cut back to within four to six inches of the bud inserted in the Pear wood. This small section or piece of wood of a vigorous-growing variety, uniting with the Quince stock below, and supporting another above, will have a very great influence in the future growth of the tree. Doubled worked trees cost twice as much to raise as single worked, but they are worth the difference to any one who desires dwarf Bartlett, Seckle, and similar varieties, which do not usually succeed when grown directly on Quince stocks. Doubled worked trees, however, should never be trained as standards, as the Quince has a rather feeble root system, producing few large lateral roots ; consequently, if the top is trained high the tree is likely to be blown over, and all of the roots thrown out.

The White Thorn (*Cratægus coccinea*), may also be employed as a stock for the Pear, and in moist, heavy soils it is fully equal, it not superior, to the Quince. There are several native species of the Thorn, some much stronger growers than others; the largest and most vigorous species should be selected, if to be employed as stocks for the Pear, or even for propagating the ornamental varieties of the European Hawthorn.

The common Juneberry or Shadbush (*Amelanchier*), and the Mountain Ash (*Pirus Americana* and *P. Aucuparia*), may be employed as stocks for the Pear, but are usually considered inferior to the Pear and Quince. The common Apple may also be used as a stock for the Pear, but the union between the wood of the two trees is very imperfect, and seldom of long duration.

PERSIMMON (*Diospyros*).—The Date Plum of Europe, and the many cultivated varieties of the Kaki, or Japan Persimmon (*D. Kaki*), when grown on the American Persimmon, thrive equally as well, if not better, than on any other stock. The American Persimmon is not only the largest, but most hardy tree of the genus, consequently preferable as a stock for the less robust species and varieties. The seedlings grow rapidly, and often reach a size large enough for budding the first season. If not used at this time, they may be headed back the following spring to insure a vigorous growth of young wood, into which buds may be inserted later in the season.

PLUM (*Prunus*).—Seedlings of the hardiest and most vigorous growing varieties of the European Plum are usually preferred by nurserymen for stocks, but the larger growing species of the native Plum answer the purpose well, if budded low, or the cions are splice grafted on the roots, and then planted so deep that the point of junction is covered. The Myrobalan Plum is a favorite among the French nurserymen and orchardists, as it is said not to produce suckers, or, at least, very sparingly, and it is readily propagated by cuttings. There are several other species or varieties of the Plum that may be readily propagated by cuttings, layers, mound layers made by banking of the sprouts or old stools, and by cuttings of the roots. The latter mode is objectionable, as stocks produced in this way are very likely to produce suckers far more freely than will be desirable as the trees become old. Varieties of the Plum, known as the Black Damas and St. Julien, are also largely used as stocks by the French nurserymen, and are also imported by those of this country. These, with the Myrobalan, are among the best, if not the very best, of the European varieties for stocks upon which to work the various cultivated sorts of the Plum, Apricot, and, I may add, the Peach, whenever it is desirable to raise the latter on any other than its own stock.

Our indigenous species of the Plum, especially the *Prunus Americana* and *P. Chicasa*, are the best stocks on which to grow all the varieties originating from the same, but the more rapid and stronger-growing European varities soon overgrow the American stocks, unless worked very low down, or directly on the roots.

The Peach is often employed as a stock for the Plum, and in mild climates, and where the trees are planted on rather light soils, it answers the purpose moderately well; but it is not recommended for general use, and in cold climates, or where the Peach is subject to diseases and attacks of insects, it is useless as a stock for the Plum.

POMEGRANATE (*Punica Granatum*).—The Pomegranate is usually propagated by seeds, cuttings, and layers, but it may be grafted; the Sweet-fruited, Violet, and other varieties being worked on the stocks of the Wild Pomegranate, or one variety may be employed as a stock for any other, whenever such mode of propagation is desirable.

QUINCE (*Pirus, Cydonia* of some).—Strong kinds, like the Angers and Fontenay, are used as stocks for the improved varieties cultivated for their fruit. Also for the ornamental, like the Japan Quince, or *Pirus Japonica*, of nurserymen's catalogues, and the large Chinese Quince (*Pirus Chinensis*). Some European authorities recommend the Portugal Quince (*P. Lusitanica*), as a stock not only for the Pear, but for other varities of the Quince, as it is a very vigorous grower.

TREE AND SHRUB STOCKS.

In making a list of the stocks usually employed in the propagation of ligneous plants, I shall not attempt to include in it every species and variety that has been, or may be under certain circumstances, employed for such purposes, but merely refer to the best likely to be available. In some families, a single species may answer, or have been proved to be the best stock for all the species of an entire genus, while in the wood of others, as has already been noted among the Cherries, there may not be the least affinity. The genus *Prunus*, sub-section *Cerasus*, is not an exceptional instance of this kind, for in all of the larger genera of trees and shrubs there are groups of species which have a general botanical resemblance, and still the wood of the species of the different groups have little or no affinity. On the contrary, there are instances where a single species of a genus will not only answer well as a stock for all the different species, but also for those belonging to different genera of the same family.

To avoid repetition and frequent reference to some particular variety or species of stock in succeeding chapters, those most usually employed in propagating ligneous plants are named here and as follows :

ABIES (The Fir).—See Coniferæ.

ABUTILON.—Almost any of the strong upright growing species will answer well as stocks upon which to work the feeble or the trailing species, like *A. Megapotamicum*, when standard plants with drooping heads are desired.

ACER (The Maple).—In seeking a stock for use in propagating any variety of a species of Maple, it is a good rule to select seedlings of its parent for stocks, except when a stronger and more vigorous species can be utilized for this purpose. The so-called Soft Maples (*A. dasycarpum* and *A. rubrum*), also known as the Silver and Red Maples, have given us a few very desirable varieties, and these succeed best when grown on Silver Maple stocks. This stock may also be employed in propagating nearly all of varieties of European Maples, but in a few instances, as with the varieties of the Sycamore Maple (*A. Psuedo-Pla-*

tanus), and Field Maple (*A. campestre*), have a stronger affinity for their own species than for others. The Japan Maples, and there are numerous varieties, succeed only when worked on stocks of closely allied species. No American or European species has as yet been tried that makes even a passably good stock for these Maples, but as seedlings of the indigenous species of Japan can now be imported quite safely and cheaply, propagators are not seriously incommoded by the failure of other species to furnish a good stock.

Æsculus (Horsechestnut).—Varieties of the European Horsechestnut (*Æ. Hippocastanum*), should be grafted upon seedlings of the species. Our indigeneous varieties and species, some of which are mere shrubs, may be grafted upon seedlings of the Buckeye (*Æ. glabra*), the largest growing native species of the genus.

Alnus (Alder).—The Heart-shaped Leaved Alder (*A. cordifolia*), makes an excellent stock for trees to be planted in a dry soil, while the Sticky Alder (*A. glutinosa*), is the best for moderately dry soils. These are both natives of Europe, but it is quite probable that some of our North American species will yet prove equally as valuable for stocks, especially those which grow to a large size, like the Red Alder (*A. rubra*), of the northwest, and the Oblong-leaved Alder (*A. oblongifolia*), of New Mexico and the regions westward.

Auraucaria.—See Coniferæ.

Arbor Vitæ.—See Coniferæ.

Arbutus (Strawberry Tree).—The common European species (*A. Unedo*) is usually employed as a stock for the various species and varieties of the genus.

Azalea.—For all the various species and varieties of the deciduous Azaleas the two North American species, known as *A. viscosa* and *A. nudiflora*, have proved to be superior, if not the very best, stocks known. They are extremely hardy, and the varieties of the Pontic Azaleas (*A. Pontica*), and the semi-evergreen of the Chinese (*A. Sinensis* or *mollis* of some authors), take readily to these stocks, forming a close and lasting union and insuring a vigorous growing plant. The American Flame-colored Azalea (*A. calendulacea*), is also a good stock, but is not quite as hardy as the first two. *A. viscosa* is the stronger grower of the two, and for this reason is usually preferred as a stock for the larger growing varieties of *Pontica, mollis* and the various hybrids between the Pontic and American species. For cultivation in our Northern States, as well as in the northern countries of Europe, the two first-named species of Azaleas are without doubt the best for stocks.

BETULA (Birch).—Seedlings of the strongest growing species, like the White Birch of Europe (*B. alba*) or the American Sweet Birch (*B. lenta*) and Paper Birch (*B. papyracea*), are preferred as stocks for the many varieties of different species now in cultivation.

CALOPHACA (Lentil Shrub).—The common Laburnum (*L. vulgare*), is employed as a stock for the *C. Wolgarica*, a low-growing shrub, native of Siberia. Grafting is only practised for the purpose of producing graceful little trees of only a few feet in height; consequently, the stocks should be tall and straight, and the cion inserted six feet or more from the ground.

CAMELLIA.—Seedlings or plants raised from cuttings of the single flowered variety are usually preferred for stocks. Double flowering varieties may, however, be re-grafted wherever desired, thereby making what are called "double worked" plants.

CARAGANA (Siberian Pea Tree).—Seedlings of *C. arborescens* are the best stocks on which to work the smaller growing species and varieties. The weeping or pendulous variety should be worked on rather tall stocks, to allow room for the growth of the drooping twigs and branches.

CARPINUS (Hornbeam).—Seedlings of the American (*C. Americana*), and European (*C. Betulus*), are used as stocks for the Cut-leaved, Oak-leaved and other varieties of the Hornbeam in cultivation. The American species is the most rapid growing tree while young, but it is said that it does not reach quite so large a size in old age.

CASTANEA (Chestnut).—For cultivation in this country the American Sweet Chestnut (*C. vesca var. Americana*), is undoubteldy the best stock for all the European and oriental species and varieties, including the recently introduced Japan Chestnut. The common American Chestnut tree grows to a very large size and is perfectly hardy where many of the varieties of the European species are tender. It is said that the Chestnut succeeds moderately well when grafted on the Oak, but it will seldom be necessary to employ Oak stocks, as seedlings of the Chestnut are usually cheap and readily obtained.

CATALPA (Indian Bean).—Seedlings or root cuttings of the common American Catalpas (*C. bignonioides* and *C. speciosa*), may be used as stocks for the less vigorous growing varieties or oriental species.

CEDRUS (Cedar).—See Coniferæ.

CERASUS (Cherry), Sub-section of *Prunus*.—The same kind of stocks are usually employed for the ornamental species and varieties of the Cherry as for those cultivated for their fruit. The dwarf and trailing varieties, when worked on tall, straight stocks of the Mazzard, form handsome, round-headed or broad-spreading, pendulous-branched trees, much admired for ornamental purposes. With some of the Japanese varieties it is best to graft or bud low down and then train to a stake until the stem reaches the desired height; then cut off and allow the head to form as though grafted high. This treatment is recommended because the wood of some of the oriental varieties is not so likely to be injured by cold in winter or by the heat in summer, as that of the Mazzard Cherry.

Varieties of the evergreen species of the Cherry, such as
C. ilicifolia, C. Lauro-cerasus and the *C. Lusitanica*,
should, of course, be grafted on stocks of their own or
closely allied species.

CHAMÆCYPARIS (Cypress).—See Coniferæ.

CHIONANTHUS (White Fringe).—Almost any species
of the Ash (*Fraxinus*), makes a good stock for the
American White Fringe (*C. Virginica*) and the Chinese
species (*C. retusus*). The European Ash (*F. excelsior*),
is, however, usually preferred to the American species as
a stock for the fringe trees.

CONIFERÆ (Cone Bearing).—With but few exceptions,
the Conifers are evergreen trees or shrubs. The evergreen kinds must necessarily be confined to stocks of the
same group and the deciduous to their own. As a rule,
in seeking stocks for the Conifers, the nearer we can keep
to the species from which the varieties under propagation
originated, the better, although in a few instances some
closely allied species may have proved to be superior for
this purpose than the original. For the *Abies* or Firs,
the European Silver Fir (*A. pectinata*), has been most
extensively used as a stock for the different species and
varieties of the genus, mainly because it was most common and readily obtained. Any of the other larger growing species native of cool climates will, however, answer
equally well.

With the Pines (*Pinus*), the species with two and three
leaves in a bundle should be employed for varieties of the
same, such as *P. sylvestris, P. s. nana, P. Mugho
compacta, P. Pyrenaica, P. densiflora*, etc. The common Austrian Pine (*P. Austriaca*), may be used as
a stock for our Western Pines (*P. ponderosa, P. Coulteri*
and *P. Sabiniana*), as these all have heavy, coarse-grained
wood, and are closely allied to the Austrian Pine. But
a good, rapid and free growing three-leaved species is

usually preferred as a stock upon which to work both the two-leaved and the three-leaved species. The common Red or Norway Pine (*P. resinosa*), is one of the very best of the two-leaved species as a stock for other closely allied species and varieties. The common White Pine (*P. Strobus*), is the best stock for all of the five-leaved species, such as *P. flexilis, P. excelsa, P. Cembra, P. Mandshurica*, etc., etc.

The Junipers (*Juniperus*), are rarely propagated by grafting, as they are readily multiplied by seeds and cuttings; but almost any of the strong-growing species, like *J. Virginiana*, will make good stocks for the varieties of feeble and strong-growing kinds. For the Arbor Vitæs, including the *Biotas, Thujas* and *Retinisporas*, the common American species (*Thuja occidentalis*), may be employed in preference to any other. For the *Piceas* or Spruces, the commom Norway Spruce is one of the best for stocks, as it is a very free-growing, hardy tree, and thrives in a great variety of soils. The true Cedars (*Cedrus*), such as the Cedar of Lebanon (*C. Libani*) and Deodar Cedar (*C. Deodara*), may be grafted on seedlings of their own species, or on those of the Mt. Atlas Cedar (*C. Atlantica*). European nurserymen prefer the latter when they can be obtained. For the Hemlock Spruces or *Tsugas*, the common North American species (*T. Canadensis*), is probably the best for stocks, it being one of the most hardy of the genus. The Larches (*Larix*), including the False Larch (*Pseudolarix*); are grafted on stocks of the common Larch, the European species (*L. Europæa*), being usually employed for this purpose. The oriental Cypress (*Glyptostrobus*), is so closely related to the common Cypress of our Southern States (*Taxodium distichum*), that the latter is used as a stock for the former, with its allied species and varieties.

CORNUS (Dogwood).—Seedlings of the common American Dogwood (*C. florida*), is the best stock upon which to bud or graft its own varieties, or those of other closely allied species. The herbaceous species, of course, are not propagated by these modes.

CORYLUS (Hazel or Filbert).—The Hazelnuts are rarely propagated by budding or grafting in this country, but in Europe the weak-growing and dwarf varieties are sometimes worked on the stronger. Seedlings of the common European Hazel (*C. Avellana*), it being one of the most hardy and free-growing species, is preferred as a stock. Seedlings are preferable to layers or cuttings, as they have a better root system; that is, longer and stronger side or brace-roots, as they are termed.

COTONEASTER.—The deciduous species grow freely on either Quince or Hawthorn stock, while the evergreen species are usually propagated by layers or cuttings.

CRATÆGUS (Hawthorn, White Thorn).—Seedlings of any of the strongest growing species may be employed as stocks for the double flowering and other varieties. For cultivation in this country, stocks of the indigenous species are preferable to the European.

CYTISUS (Broom, Scotch, Spanish, etc.)—The smaller, trailing, or low-growing species may be grafted on stocks of the stronger growing, or on the Laburnum, if standard plants are desired.

DAPHNE (Spurge Laurel).—The Daphnes are principally low-growing evergreen shrubs, rarely propagated by grafting, but scarce species and varieties may be grown on stocks of the common European Spurge Laurel (*D. Laureola*).

EUONYMUS (Spindle Tree, Burning Bush, Wahoo, etc.) —Seedlings of the larger and stronger growing species, like the American Burning Bush (*E. atropurpureus*), and the European Spindle tree (*E. Europæus*), are often employed as stocks upon which to work the broad-leaved species (*E. latifolius*). The evergreen species are usually propagated by layers and cuttings.

Exochorda (Great-Flowered Spiræa).—No suitable stock has as yet been found for these noble shrubs from the North of China, but small pieces of their own roots are employed in place of larger stocks, the cion being affixed to them by splice grafting.

Fagus (The Beech).—Seedlings of the American Beech (*F. ferruginea*), and the European (*F. sylvatica*), are generally employed as stocks for the different varieties in cultivation. Varieties of the Evergreen Beeches of South America and New Zealand would, of course, be grown on stocks of the species from which they originated.

Fraxinus (Ash).—Seedlings of the common European Ash (*F. excelsior*), are the best for stocks for all the European and American species and varieties. Long experience in the use of the European Ash as a stock for the many cultivated varieties, has fully established its reputation as one of the best, if not the very best, species to be employed as a stock. The young wood is soft, fine grained, and either buds or cions unite to it readily. The different species of the American Ash may, of course, be employed as stocks for their own or foreign varieties, but the European Ash is usually preferred.

Gleditschia (Honey Locust).—The common American Honey Locust, or Three-Thorned Acacia (*G. triacanthos*), is an excellent stock for the Chinese species and varieties, as well as the thornless and other varieties of our native species.

Halesia (Snowdrop Tree). — As the Four-winged Halesia (*H. tetraptera*), is the largest growing and most hardy species of the genus, it is the best stock. This, and closely allied species, may also be employed as stocks for other species of the *Styracaceæ* or Storax family, as, for instance, the Japan Styrax (*S. Japonica*), and

Pterostyrax hispidum, and the different North American species of Styrax.

HALIMODENDRON (Silver-Leaf). — These Siberian shrubs succeed best when grafted on stocks of the *Caragana arborea*, or Siberian Pea Tree.

ILEX (Holly).—The common evergreen American Holly (*I. opaca*), is without doubt the best stock for the closely allied species and varieties, especially if they are to be cultivated in cool or cold climates. But as seedlings of this species are not so readily procured as those of the European Holly (*I. Aquifolium*), the latter are more generally used for this purpose.

JUGLANS (Walnut, Butternut).—Seedlings of the larger growing varieties of the European Walnut (*Juglans regia*), are usually employed as stocks for the different cultivated varities. It is quite probable that the common American Butternut (*J. cinerea*), could also be utilized for the same purpose, but further experiments are needed to determine its real value as a stock.

LARIX.—See Coniferæ.

MAGNOLIA.—Seedlings of the common American Cucumber-Tree (*M. acuminata*), are usually recommended as the best stocks for all of the deciduous species of the Magnolia, whether natives of China, Japan or America. But Mr. J. R. Trumpy, of the Kissena Nurseries, of Flushing, N. Y., who has probably propagated a greater number of species and varieties of the Magnolia than any other man in this country or Europe, is quite emphatic in giving the preference to the Umbrella Magnolia (*M. Umbrella*), also called *M. tripetala*. He says that the latter species is more easily worked and produces a greater number of fibrous roots, consequently is not so seriously affected by transplanting.

PLANERA (Planer Tree).—All the species and varieties from Japan and Siberia, as well as those indigenous to

North America, grow freely on the Elm (*Ulmus*). The common American and English Elm may both be employed as stocks for the Planeras.

QUERCUS (Oak).—In selecting stocks for the Oaks, the propagator will secure the best results by taking seedlings of closely related species of each of the several groups that are usually designated under such names as White Oaks, Black and Red Oaks, Chestnut Oaks, Willow Oaks and Evergreen Oaks. The English Oak and our native White Oaks are closely allied and may be interchanged as stocks for each other. The European Oaks (*Q. Robur* and *Q. pedunculata*), will probably serve as stocks for a larger number of species and varieties than any other two species that have been tried. Most of our North American Oaks take on these quite readily, while, on the contrary, very few of the American species will answer as stocks for the European varieties. The Chestnut Oaks come next, as the acorns of both groups mature the first year. The Willow Oaks are biennial fruited, and some of them almost evergreen, consequently not so closely allied as the two first groups. The Scrub, Black and Red Oaks have rather coarse-grained wood, and are rather indifferent stocks to work, even for varieties of their own species. The *Q. Ilex* is the species most usually employed as a stock for the evergreen species and varieties, although most of the evergreen Oaks may be readily propagated by cuttings.

RHODODENDRON (Rose Bay).—The *R. Ponticum*, from Southern Europe, has been more extensively employed as a stock than any other species, and while it has served the purpose well in Europe, it is inferior in growth and hardiness to our native species, *R. maximum* and *R. Catawbiense*. The former is superior to the Pontic species as a stock for the Rhododendron in this country, although probably not so readily obtained, or so cheap.

ROBINIA (Locust or False Acacia).—Seedlings of the common Locust (*R. Pseud acacia*), are the best stocks for the varieties of the species, as well as those of the Rose Acacia (*R. hispida*). The latter and its varieties, when grown on their own roots, produce suckers so freely that they become a nuisance in the garden.

ROSA (Rose). — Many species and varieties are employed as stocks, and in Europe the Wild Dog Rose (*R. canina*), is usually recommended for this purpose. The wild plants as they are found in the hedges and woods are used, as well as seedlings raised in the nursery. A variety known as the Manetti is extensively employed as a stock in this country, and appears to thrive better in our climate than the Dog Rose, and for this reason is usually preferred. It is readily raised by cuttings, these producing plants large enough for use the first season. The common Sweet Briar (*R. rubiginosa*), which is a naturalized species from Europe, is also an excellent stock for nearly all of the cultivated varieties. It is less likely to produce suckers than the Manetti, and is exceedingly hardy.

SALIX (Willow).—The varieties usually propagated by grafting will, as a rule, succeed best on stocks of their own species. The common Kilmarnock Willow being only a variety of the English Goat Willow (*S. Caprea*), the cions take more readily on this stock than on any other. It will grow, however, on the Pointed-leaved Willow (*S. acuminata*), also indigenous to Great Britain, but the cions do not take as readily and freely, but when they do unite, the union is quite perfect and lasting. The so-called American Weeping or Fountain Willow (*S. purpurea pendula*), is a trailing variety of the English Bitter Willow, and may be grafted on stocks of either of the above-named species.

SOPHORA JAPONICA.—The Weeping and Variegated-leaved varieties are grown on seedling stocks of the species.

Tilia (Linden or Basswood).—The common American Linden (*T. Americana*), is one of the largest and most rapid growing species of the genus. It is also less liable to the attacks of insects and diseases than the European species; consequently, to be preferred as a stock for all the different varieties and species as yet brought to notice. Seedlings are better than plants raised from cuttings or layers.

Ulmus (Elm).—The varieties of the different species succeed best on the mother stock. That is, the many varieties of the English Elm (*U. campestris*), should be grown on the stocks of the original species, and those of the Scotch or Wych Elm (*U. montana*), on their own species. The common Camperdown Weeping Elm belongs to the last species, and should be grafted on seedling stocks of the same. Of course, in case the proper stocks are not at hand, other closely allied species may be used, but it is always best to select stocks from the species from which the varieties originated.

CHAPTER XIX.

INFLUENCE OF CION AND STOCK.

In selecting cuttings, cions and buds, it is well to keep in mind the fact that they have more or less influence in determining the future value of the plant raised therefrom. Whatever faults or merits are possessed by the parent plant are likely to be transmitted to the offspring, and either, under certain conditions, may be increased or decreased many fold. If we desire early fruiting, we should select wood for propagation from mature or bearing plants, instead of from the young and immature. But we may readily carry this kind of selection too far, for very early and premature fruiting is not always desirable, especially with trees which need to be of considerable size, to enable them to sustain a fair or large crop. Continuous propagation from old, mature and productive specimen trees may increase the tendency to a premature old age and decay. Almost any peculiar form of growth, or other characteristic of a variety or species, may be transmitted to the offspring through the part employed in its propagation. We may not in every instance be able to perpetuate abnormal characteristics at first, but by repeated selections of parts showing a variation from the normal type, we can usually fix and perpetuate almost any peculiar habit or form of plant.

In selecting parts of herbaceous plants, the same rule holds good as in the ligneous, and we may increase the floriferous habit by continuous propagating from the flowering stems and branches, until the plant perishes from what may be termed over-exhaustion.

INFLUENCE OF STOCK ON CION.—That the stock upon which a cion is set has influence upon its future growth is well known. If it were not so, then the art of propagating plants by budding and grafting would be less valuable than now.

The stock not only acts as a medium through which the cion obtains sustenance from the earth, but it in a measure imparts its own characteristics to it; and it is thus we change the giant into a dwarf, the slow growing plant into a rapid one, and many other variations from the natural habits of plants, simply through the influence of the stock on the cion or graft.

While we may not, in every instance, be able to determine the true cause of certain variations, which may appear to be antagonistic with what we call natural laws, still, for all practical purposes, our knowledge of this subject is sufficient to enable us, in many instances, to so change natural products that their value is increased many fold.

The common mode of producing dwarf trees is one of the most familiar instances of the influence of the stock on the graft. It mainly affects the form and habit of growth, but is not necessarily debilitating, for size and rapid growth are not always trustworthy signs of perfect health or great longevity.

In some instances we employ stocks as a mere temporary support to the cion, not expecting or desiring a permanent union, as in grafting the tree Pæonia upon the tuber of the herbaceous, or the stem of one Dahlia upon the tuber of another. But with trees we usually seek permanency, and therefore select stocks that shall not only support the graft, but aid in developing those particular characteristics which are most desired. The influence of the stock upon the graft may be briefly stated as follows :

First, The stock gathers the crude materials for the support of the graft from the soil, and in doing so it may supply it in such quantities as to produce rapid growth, or the reverse.

Second, The tendency of the stock is to impart its own habit of growth to the graft.

Third, One species of stock will extract from the soil the peculiar components which are necessary to support the graft, while another will not ; consequently, a variety or species may fail upon one stock and succeed upon another in the same soil and locality.

Fourth, The hardiness of a tree is but slightly changed or affected by the stock, except as its growth is influenced to mature early or late in the season.

Fifth, The quality of a fruit is occasionally influenced by the stock, but the true cause of this is not as yet sufficiently understood to allow of any rules being given by which it may be avoided. Size of fruit is also in some instances considerably changed by the use of different stocks. I have known two Bartlett Pear trees of the same age, standing side by side, and apparently of equal vigor, still, for ten years, one has produced very large fruit and the other small. The number of specimens upon each tree being reduced to an equal number, the difference in size remained the same. With such examples before us, we cannot but conclude that the stock, in some instances, does exert sufficient influence to change the size of the fruit, as well as the form of the tree.

Sixth, The stock will not only impart vigor to the graft, but also transmit diseases. It is therefore just as important to avoid the one as to endeavor to secure the other.

INFLUENCE OF THE CION ON THE STOCK.—The influence of the cion on the stock is a subject only occasionally referred to in our modern horticultural works. Downing says: "The influence of the graft on the stock seems scarcely to extend beyond the power of communicating disease." But, if we have discovered this much, it proves that there is an influence, and if it is sufficiently potent to "communicate disease," then it is probably powerful enough to impart other properties as well. Mr. J. J. Thomas, in his "American Fruit Culturist," edition of 1849, says: "The extension of the stock by successive depositions from the leaves of the graft and through the cellular system of the bark, so as to preserve the strict specific identity of the wood of the former, is familiar to every practical cultivator." The same seedling Cherry stocks, grafted with sorts of different degrees of vigor, soon vary in amount and size of the fibrous roots. Trees of the Imperial Gage and Jefferson Plums, a few feet in height, when budded on the Wild Plum, were found to have only half the amount of roots possessed by the unbudded stock of the same age.

Every nurseryman must have observed that some varieties of the Pear, as well as of the Plum and Cherry, have a far greater number of fibrous roots than others. So marked is this difference that the common laborers in the nursery soon learn to distinguish them and will proceed quite differently in digging the trees of each variety, knowing that one has few long naked roots, while the others have short and numerous fibrous ones. These various forms of roots cannot be satisfactorily accounted for in any other way but to ascribe the cause to the influence of the graft. If we take a seedling Apple tree of one or two years old, and divide the root into two parts, upon one of which we splice a cion of Monmouth Pippin, and on the other one of the Northern Spy, and plant both in exactly the same soil, side by side, and cultivate them alike, after three or four years the roots will have a decidedly different appearance both in color and form. Still, with all the influence the cion has had upon the roots in changing their form and color, if cuttings are taken from these roots and forced to produce shoots, the plants thus raised will be of the original type, showing that the influence of the cion is not perpetual, but continues only so long as the roots are in position to gather the crude nutrients from the soil for the leaves on the cion to assimilate; thus, while this reciprocal action continues, whether it be for one or fifty years, the cion will continue to hold its influence over the stock or roots.

A few instances have been recorded where the cions with variegated leaves have so influenced the stock as to cause it to produce shoots below the point of union, bearing leaves like those on the cion. But whether this change is due to some disease inherited in the cion, or the intermingling of the cellular matter of the two parts, has never been fully determined. Although this subject of reciprocal action between stock and graft has been frequently referred to by writers on horticultural topics, from the time Pliny wrote his "Historia Naturalis," down to the present, still, there does not appear to have been any very carefully conducted experiments made for the express purpose of ascertaining its exact extent or limits. It remains an almost wholly unexplored field, to be occupied by some future disciple of vegetable phenomenology.

CHAPTER XIX.

SELECT LISTS OF PLANTS.

In the following lists of plants annuals are omitted, for it is presumed that every cultivator of the soil knows that these are generally propagated by seed. There may be an occasional instance where it is desirable to perpetuate an annual by cuttings, but these may be considered as exceptions to a general rule. Furthermore, as this work is not intended to be a botanical dictionary, nor an encyclopædia of plants, the author only aims to mention those species and varieties which are to be found in cultivation either in the gardens or conservatories of the inhabitants of temperate climates.

The plants will be named in alphabetical order according to their botanical names, one or more of the common or local names being added when known; but there are many species to which none have as yet been given—a fact not at all to be regretted, as local and popular names are usually as untrustworthy as they are unnecessary and confusing.

TREES, SHRUBS AND VINES, WITH BRIEF NOTES ON HABITS AND MODES OF PROPAGATION.

Abelia.—Mostly evergreen shrubs, adapted to cool greenhouse culture in cold climates. Propagated by green cuttings taken off in summer and planted in a close frame, or by layering in the house.

Abies (Balsam Fir, Etc.)—Well known coniferous trees, propagated by seeds preserved dry over winter, and then sown in light soil in frames, or where water and shade can be applied when required. Varieties are propagated by veneer grafting under glass late in summer.

Abroma.—East India and New Holland evergreen trees; succeeding only where they can be given a high temperature. Increased by seeds or cuttings of the half-ripened wood placed in a close frame or under a bell glass.

Abutilon ("Flowering Maples").—Very free blooming ornamental trees and shrubs, natives of warm climates, but succeeding in a cool greenhouse in winter and in the garden during the summer. New varieties are raised from seed, then increased by cuttings of the young shoots planted in sand in frames. Cuttings root very freely, and usually without dropping their flower buds if these are left on when the cuttings are made.

Acacia.—Evergreen trees and shrubs, principally from the tropics—Australia, New Holland, East Indies, South America and Africa. About 400 species are known, but not more than one-eighth of the number in cultivation. Propagated by seed, or cuttings taken off with a heel and inserted in sand under a bell glass or in a close frame. Also by root-cuttings of two or three inches in length, placed in sand, with the larger end only lightly covered.

Acalypha.—Tropical shrubs with inconspicuous flowers, but rather ornamental foliage. They require a high temperature to bring out the bright color of their leaves. Propagated by cuttings, taken off in early spring and placed in a close frame, and given a temperature of eighty to ninety degrees, Fahrenheit.

Acer (Maples).—Well-known deciduous trees and shrubs, natives of cool climates. Seeds of such species as American Silver-leaved Maple (*A. dasycarpum*) and Red Maple (*A. rubrum*), which ripen early in summer, should be sown immediately and covered very lightly; but the seeds of species ripening in autumn may be readily preserved by mixing with clean sand and then stored in a cool place until spring. These late ripening seeds may also be sown in autumn if preferred, although there is greater danger of loss from vermin than when stored in sand over winter. The varieties of the species named may be readily propagated by budding or grafting in the nursery, and the same is true of the Hard or Sugar Maple; but in grafting, the cions should be taken from the trees early in winter and kept dormant until the sap has begun to flow quite freely in the stocks. The European varieties require similar treatment; but the Japan Maples are more successfully grafted under glass and by veneer grafting, the stocks having been grown in pots for this purpose. The Japan Maples may also be propagated by cuttings of the green wood taken off in summer, but the plants are usually feeble, making a very slow growth; consequently, this mode of propagation is not recommended. The Negundo Maple or Box Elder (*A. Negundo*), may be propagated by ripe wood cuttings taken in the fall and placed in a moist and warm position, where a callus will be formed by the time they are wanted for planting out in the spring. All the species and varieties of the Maple may be increased by layers, made in autumn or after the leaves are nearly full formed in spring; but plants raised from layers are inferior to those produced from seeds, or by budding and grafting.

Actinidia (Japan Gooseberry).—A small genus of hardy deciduous climbing shrubs from Japan, one of the species bearing edible berries, resembling a gooseberry in size and flavor. Propagated by seeds, layers and cuttings of the green shoots in summer. Both cuttings and layers produce roots freely, but the buds on them push very slowly and often fail, although the young plants may have an abundance of roots.

Adenocalymna.—Evergreen climbing shrubs, belonging to the same order as the common *Bignonia* (Trumpet-creeper), but being native of a tropical climate, they require great heat to insure vigorous growth and perfect flowers. Increased by cuttings placed where they will receive plenty of moisture and bottom heat.

Adenocarpus.—A genus containing both evergreen and deciduous shrubs, bearing long racemes of yellow pea-shaped flowers. Propagated by seeds, layers and cuttings of the unripe wood under glass.

Adenostoma (Chamiso).—A genus of only two species of small evergreen trees or shrubs indigenous to California. Propagated by cuttings of the immature shoots in sand under a bell glass or in frames

Adhatoda. — Greenhouse or stove evergreen shrubs from Brazil and India. Propagated by cuttings of the young shoots placed in a position where they will receive bottom heat.

Ægiceras (Goat Plant). — One species from New Holland, sometimes cultivated for its white, fragrant flowers. A rather stocky shrub. Propagated by cuttings of the half-ripened shoots.

Æsculus (Horsechestnut).—Deciduous trees or shrubs. Numerous species and varieties in cultivation. The species are usually propagated from seed. The large, fleshy nuts should be gathered as soon as they fall from the trees in autumn—the

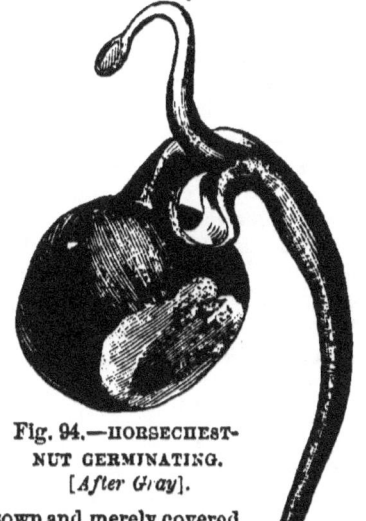

Fig. 94.—HORSECHEST-NUT GERMINATING. [*After Gray*].

outside husk removed—then either sown and merely covered with leaves or very light vegetable mold, or the nuts may be preserved in pure sand or sphagnum until the following spring. They should be stored in a cool place to prevent premature sprouting and decay. If the boxes containing the nuts are buried in the ground under an evergreen tree, or in the shade of some building, they will usually pass through the winter in good order. In the germination of the Horsechestnut, the nut does not burst open as in the Acorn, Butternut and Hazelnut, but both plumule and radicle press through the envelope on one side, as shown in figure 94. Varieties of the different species are propagated by budding and grafting. The most certain mode is by veneer grafting under glass, and in this climate during the month of August. Seedling stocks, placed in pots in early spring, may be grafted the first summer, as the stocks will produce plenty of small fibrous roots by the time they are wanted for use. Some of the dwarf species, like the California Buckeye (*Æ. Californica*), and the Dwarf Buckeye (*Æ. parviflora*), of the Eastern States, are readily increased by dividing up the old stools or clumps of stems.

Aganosoma.—Showy greenhouse shrubs, principally from India. Propagated by cuttings under glass with gentle bottom heat.

Agapetes.—Deciduous shrubs from the mountains of India. They are closely allied to the Huckleberries (*Vacciniaceæ*), but require the heat of a warm greenhouse. Propagated by seeds, and half-ripened cuttings under glass.

Agathophyllum (Madagascar Nutmeg).—An evergreen tree closely allied to the Laurels, the leaves having the fragrance of the clove. Requires the heat of a warm greenhouse. Propagated very readily by cuttings of the green shoots.

Agathosma (Bucco).—Small, evergreen, heath-like shrubs from the Cape of Good Hope, thriving in a cool greenhouse in winter and in a half shady position during the summer. Propagated by cuttings of the green, succulent shoots under glass.

Ailantus (Tree of Heaven).—A well known tree from China; the staminate flowers exhaling a disagreeable, nauseating odor. Propagated by seeds preserved in a dry, cool place over winter and then sown in spring and lightly covered. Also increased by suckers and cuttings of the roots. The latter modes are not recommended except for propagating the pistillate trees, the flowers of which are odorless.

Akebia.—A very hardy and handsome twining shrub from Japan. Only one species as yet known, the *A. quinata*, or Five-leafletted. Readily propagated by layers of either the old or young shoots.

Alhagi (Manna Tree).—Small shrubs with pea-shaped flowers. Natives of Caucasus. One of the species, the *A. maurorum*, yields the substance known as "manna," a natural exudation of the leaves and branches. Propagated by seed, and cuttings of the green shoots placed where they will receive bottom heat.

Alnus (Alder).—Deciduous trees and shrubs, mostly natives of cold countries. The species are usually propagated by seeds preserved in a dry, cool place over winter, and sown on the surface of the soil in spring and thinly covered with moss or some light vegetable mold, which should be kept constantly moist until the plants appear. Varieties are propagated by cuttings of ripe wood, layers and grafting upon free growing stocks. The recently introduced Japan Alder (*A. firma*), succeeds best when grafted on the European Sticky Alder (*A. glutinosa*).

Amelanchier (Juneberry, Shadbush, Etc.)—Deciduous trees and shrubs. The North American species are extremely variable, producing many natural and widely different varieties. Some grow to trees thirty or more feet in height, others are merely dwarf shrubs two or three feet high; all producing edible and pleasant-tasting fruit. Propagated by seeds, layers or cuttings of the ripe roots. The dwarf varieties, when grafted on tall stocks of the larger growing kinds, form handsome, small, round-headed trees.

Amorpha (Lead Plant, Indigo Shrub).—A genus of a few species of hardy shrubs, all natives of North America. There are several local forms or varieties in cultivation. Propagated by seeds, layers, sprouts and ripe wood cuttings, taken off early in the fall and planted in a half shady position, and left undisturbed until the following autumn.

Amygdalus (Almond).—Shrubs and trees of Almond or Plum genus, or *Prunus* of most of the modern botanical works. Propagated by seeds or by grafting and budding on Almond, Peach and Plum stocks. The Dwarf Double Flowering Almond (*A. nana*), is readily propagated by cuttings of the larger roots, made in autumn and stored in sand or moss in a cool cellar until spring, then sown in drills and covered about two inches deep with light soil. These dwarf varieties may also be budded on Peach or Plum stocks, and if the buds are set three to four feet from the ground, very elegant little trees may be produced. Plum stocks are preferable to the Peach, as the latter are liable to be attacked by the Peach tree borer.

Andromeda.—Neat little shrubs—several of the species evergreen—nearly all quite hardy in our Northern States, although some are natives of the South. Propagated by seeds sown in very light soil and in seed-pans or shallow boxes, kept shaded and constantly moist until the plants appear, then removed to a position where they will receive more light. The Andromedas may also be propagated by layers; but these produce roots slowly, and it usually requires two years to secure well-rooted specimens. Nurserymen usually obtain their stocks from the native habitats of the species, as the young plants may be transplanted without much loss.

Anona (Custard Apple).—Trees and shrubs mostly tropical, some of the species bearing highly-prized and delicious fruit, like the Cherimoyer (*A. cherimolia*), Sour-sop (*A. muricata*), and Sweet-sop (*A. squamosa*), of the West Indies and South America. Propagated by seeds placed in a moist and high temperature, and by cuttings of the mature wood under glass and with bottom heat.

Aralia.—A genus of the order *Araliaceæ*, containing numerous species of trees and shrubs, and a few herbaceous plants. Among the ligneous section the American Angelica tree or Hercules Club (*A. spinosa*), is perhaps the most common in our gardens. The Chinese Aralia (*A. Chinensis*), is a closely allied species, and moderately hardy when grown in a rather dry, open soil. The recently introduced Mandchurian Aralia (*A. Mandchuricus*, or *dimorphanthus* of some authors), is as hardy as our indigenous species, and is a shrub worthy of the attention of apiarians, as it blooms profusely and its flowers yield a large amount of honey. All of the hardy, shrubby species, and some of the tender ones, are readily propagated by cuttings of the roots taken off when the plants are in a dormant state in the fall. The root cuttings should be preserved in sphagnum, or sharp sand, until spring, or the hardy species planted out in the nursery and the tender ones forced under glass. Some of the tropical species are not so easily propagated as the hardy, but they may be grafted under glass, using stocks of the more free-growing sorts for this purpose. All the species may be increased by seeds, whenever these can be procured.

Araucaria (Norfolk Island Pine, Etc.)—A genus of cone-bearing evergreen trees, natives of the Southern Hemisphere, none of the species quite hardy in our Northern States, although often cultivated in tubs and pots for decorative purposes; protection being given them during the winter. Propagation by seeds is the most satisfactory method, but all may be multiplied by layers and cuttings; the latter should be made from the ends of the shoots, and placed in sand and in a rather cool house until callused, then given a higher temperature.

Arbutus (Strawberry Tree).—Evergreen trees and shrubs. Some of the species are hardy, others require the protection of a greenhouse. Propagated by seeds sown in the fall or early in spring, and by layers; also by budding on strong seedling stocks.

Arctostaphylos (Bear Berry).—A genus of low-growing and trailing shrubs, closely allied to the last, and propagated by similar methods.

Ardisia.—A very extensive genus of evergreen trees and shrubs, all of tropical or semi-tropical origin, cultivated principally for their ornamental berries, which are quite persistent, remaining on the plant several months after they have assumed their brilliant colors. Propagated by seed sown as soon as ripe, and by cuttings of the half-ripened shoots, planted in frames or under bell glasses.

Aristolochia (Birthworts).—A genus of about one hundred and seventy species, mostly twining shrubs, the large majority being natives of tropical countries. There are a half-dozen species indigenous to the United States. The Dutchman's Pipe (*A. Sipho*), is a species in common cultivation, as it is one of the most hardy. Propagated by layers, cuttings of the roots, or of the green shoots planted under glass.

Artocarpus (Bread Fruit).—A genus of evergreen tropical trees, requiring a high temperature to insure a healthy growth and perfection of their fruit. The true Bread Fruit (*A. incisa*), is sometimes cultivated for its ornamental foliage, but the fruit seldom reaches maturity, except in the tropics. The species of this genus are all difficult to propagate under artifical conditions. Cuttings and suckers may be utilized for this purpose, but do not grow very readily or freely.

Asimina (Papaw, Custard Apple).—A genus of North American trees and shrubs of the order *Anonaceæ* or Custard Apple Family. The large Papaw (*A. triloba*), is a well-known small tree, extending from Lake Erie, in the north, to the Gulf of Mexico, in the south.

Propagated by seeds and layers put down in autumn. Seedlings usually spring up in great abundance about the wild plants, but are somewhat difficult to make live unless transplanted while young and of small size.

Athrotaxis.—A small genus of Tasmanian evergreen trees and shrubs, belonging to the *Coniferæ*. Rather tender, but will no doubt succeed in the Southern States. Propagated by seeds, when these can be obtained, otherwise by cuttings under glass.

Atragene.—See Clematis.

Aucuba (Golddust Tree).—Evergreen deciduous shrubs from Japan and the Himalayas. The pistillate plants bear very showy fruit or berries, but these are not usually obtained in the absence of artificial fertilization of the flowers. Where ornamental berries is the special object in the cultivation of these plants, specimens of both sexes should be grown in the same house, and the pistillate flowers carefully fertilized with pollen from the staminate, applying it, as usual, with a fine camel's-hair pencil. All the varieties are readily propagated by cuttings of the green or half-ripened wood, planted in sand, or almost any kind of light soil.

Azalea.—A genus of evergreen and deciduous shrubs, all very ornamental and exceedingly popular for both greenhouse and garden culture. The North American and European species have been hybridized, and from these hybrids an immense number of varieties produced, many of which are far superior to any of the parent species. These hybrids, and the seedlings therefrom, are known under the popular name of "Ghent Azaleas." The Chinese Azaleas (*A. sinensis*), from both China and Japan, are slowly deciduous, the foliage remaining on the plants until late in the autumn, but all are nearly, or quite, hardy in our Northern States. Of this species there are a large number of varieties cultivated in Japan, and recently introduced into our gardens under the name of *A. Mollis* or soft-leaved. The Indian species (*A. indica*), are evergreen and usually tender, although an occasional variety may survive in the open air if given a little protection in winter; they are generally cultivated under glass, but do not require a very high temperature. The evergreen varieties are propagated by seeds and cuttings of the young shoots, taken off with a heel or close to the old wood, and then placed in sand, and in a close frame in the house. The deciduous varieties are propagated by layers, divisions and by veneer graftings in summer under glass. (See Selecting Stocks, Chapter XVIII.)

Azara.—A genus of graceful half-hardy shrubs from South America. The flowers are mostly yellow, with an aromatic fragrance. Propagated by ripened cuttings placed in moderate heat, and under glass in cool climates, and in simple frames without artifical heat in warm climates.

Baccharis (Groundsel Tree).—A genus containing shrubs, trees, and herbaceous plants, but none of any special value or interest to cultivators of plants. There are two shrubby species found along our coast, from Connecticut southward, the most common being known as the Groundsel tree (*B. halimifolia*). Propagated by seed and ripe wood cuttings.

Banksia.—A genus of evergreen shrubs, native of Australia, and cultivated for the beauty of their foliage. There are a large number of species cultivated in European gardens, where they are employed for table decorations and for ornamenting rooms on festive occasions. Propagated by well-ripened cuttings, separated carefully below a joint, and then planted in sand without removing any but the lower leaves. Only moderate heat is required, and the air in the frames should not be too confined or moist.

Benthamia. — A genus of shrubs, of the order *Cornaceæ*, and by botanists it is now referred to as the genus *Cornus* (Dogwood). Of the two species in cultivation, the *B. Japonica* is the most hardy, but the leading shoots suffer, more or less, every winter in my grounds. Propagated by seeds and layers, or by grafting on the Dogwood.

Berberidopsis. — An evergreen, half-climbing shrub from Chili; closely related to the common Barberry. Propagated by seeds, green cuttings, and layers of mature shoots and branches

Berberis (Barberry).—A genus of many species, mostly evergreen, erect or trailing shrubs. A few species are deciduous, like the common European Barberry (*B. vulgaris*), and the American (*B. Canadensis*).

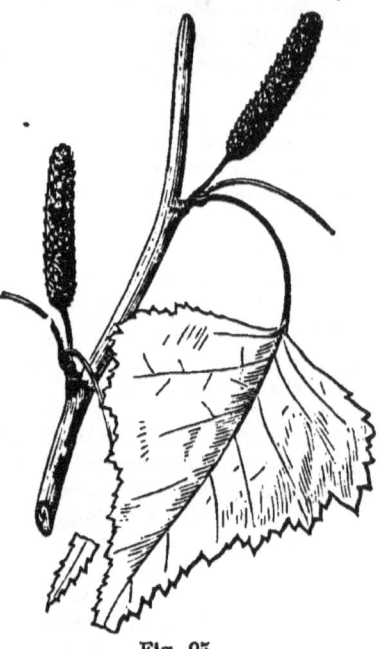

Fig. 95.

WHITE BIRCH LEAF AND CATKINS.

Increased by seeds, by cuttings of the mature wood in autumn, by layers, and some of the evergreen species by cuttings of the subterranean branches. The seeds of the species indigenous to cool climates should be washed from the pulp, mixed with sand and buried in the ground out-doors over winter, and sown in a half shady place in spring, as the young plants are very sensitive to the direct rays of the sun when they first appear above ground. (See *Ilex*).

Betula (Birch).—A genus of about thirty species of graceful deciduous trees and shrubs. Fully one-third of the known species are found in North America, and some of them extending far northward. Flowers appear before or with the leaves—sexes in separate catkins; the males long cylindrical, as shown in figure 95 (common White Birch); the females similar, but more dense or shorter; both drooping when nearly mature. Seeds small, nut-like, winged. The species are usually propagated by seed, which ripen early in autumn, and may be kept dry over winter and sown very early in spring; or they may be preserved in moist sand and stored in a cold place to prevent fermentation. The seed-bed should be shaded, also the young seedlings, until they are a few inches high. The seedlings should be transplanted when a year old. Varieties are readily increased by budding and grafting upon seedling stocks in the nursery. (See Selecting Stocks, Chapter XVIII.)

Bignonia (Trumpet-Flower).—A very large genus of climbing shrubs, mostly natives of warm climates; consequently requiring the protection of a greenhouse when cultivated in cold ones. In most of our recent botanical works the species are separated and those with a certain form of flower are called *Tecomas*, and the other *Bignonias*. The latter is the most common name, and applied indiscriminately to the species in cultivation. Our common will Trumpet-creeper of the Middle States (*Tecoma radicans*), is a good type of the Tecomas; while the Southern evergreen Trumpet-creeper (*Bignonia capreolata*), may be taken as a type of the true Bignonias. All are readily propagated by seed, cuttings, layers, and some of the hardy species from cuttings of the larger roots. The evergreen species may be increased by cuttings of the half-ripened shoots placed in sand under a bell glass or in close frames in a greenhouse.

Borbonia.—Ornamental tender shrubs from the Cape of Good Hope, bearing handsome, pea-shaped flowers. Increased by cuttings of the half-ripened shoots taken off late in winter and planted in sand and kept in only moderate heat.

Boronia.—A genus of ornamental greenhouse shrubs, mostly from New Holland. They only require moderate heat and should be planted out in summer. Propagated by cuttings of the half-ripened wood placed in a frame and given plenty of air, with a temperature not above sixty degrees, and only sufficient water to prevent wilting.

Bouvardia.—A small genus of slender evergreen shrubs indigenous to Mexico and South America. They are extensively cultivated for cut-flowers during the winter months. There are both single and double flowering varieties and of various colors, from brilliant scarlet to the purest white. They are all rather difficult to propagate by cuttings of the shoots, but are readily and rapidly increased by cuttings of the roots. To obtain the latter, an old plant or one of good size should be taken out of the pot, the soil shaken off, and the roots divided into pieces about one inch long. These cuttings should then be placed in shallow boxes filled with sand or sandy loam, and covered about one-half inch deep. Apply water sufficient to settle the soil, and set aside for a week or two where they will receive but moderate heat while the callus and buds are forming. The boxes may then be placed in a position where the cuttings will get a little bottom heat, which will force out the buds and sprouts. When the young plants are of sufficient size to handle readily, they may be potted off, using two or three inch pots. The time to make root-cuttings is after the plants have ceased flowering, or when they are in a semi-dormant state, as they usually are during the latter part of winter. Cuttings of the young terminal and side shoots, when about two inches long, are in the best condition for striking, and if placed in pure sand in close frames, or under bell glasses, the larger proportion can be made to grow, but the plants are so easily propagated by root-cuttings that our florists depend mainly upon this mode for increasing their stock of Bouvardias.

Broussonetta (Paper Mulberry).—Only one species, of which there are several varieties in cultivation. All low-growing trees from China. Propagated by seed, cuttings of the mature wood taken off in the fall and stored during winter in a moderately warm place. Also increased by layers, suckers and cuttings of the roots.

Buxus (Box Tree).—A genus of well-known, hardy evergreen trees and shrubs. There are many varieties in cultivation, all of which thrive best in a light, well-drained soil. Propagated by seed, cuttings and layers. Green cuttings, taken from the plants in summer and placed in frames and shaded during the heat of the day, produce roots quite freely, and in a few days. Ripe wood cuttings, taken off in the fall and placed in boxes in a cool greenhouse, will usually become well rooted by the following spring. Some of the varieties grow quite readily by cuttings planted in the open ground in spring, but the soil should be packed firmly about the base of the cuttings to insure the production of roots.

Callicarpa (French Mulberry).—Handsome little shrubs, cultivated for their ornamental berries. A genus of only five species—one American, one from Japan, two from India, and one from China. Propagated by seed, divisions, or cuttings of the young shoots in spring, placed under a bell glass in a greenhouse or in close frames where they will receive a little bottom heat.

Callistemon.—Handsome greenhouse shrubs, most of the species being native of New South Wales. Propagated by seed when it can be obtained, but the best flowering plants are raised from the ripe wood—the cuttings placed in pure sand in frames or under a bell glass.

Callitris.—A small genus of tender, evergreen trees, closely allied to the Arbor-vitæs, but with very long, slender, jointed branches. Propagated by seed, and cuttings of the branches taken off in the autumn and kept in a cool greenhouse through winter, or stored in a frame where they will not freeze.

Calluna (Heather, Ling).—A genus of the Heath family, indigenous to Europe and to rather limited areas in North America. Propagated by cuttings of the green, tender shoots planted in pure sand under glass. To secure the best cuttings, the plants should be grown in the house and the cuttings taken when the new shoots are of a proper length to make cuttings two to three inches long.

Calothamnus.—Tender evergreen shrubs of the "Myrtle Family," natives of West Australia. Flowers bright scarlet. Propagated by cuttings of the young shoots, when they have become somewhat firm, placed in sand in a frame. The air should not be kept too moist, as there is danger of the cuttings damping off.

Calycanthus (Sweet Scented Shrub).—A genus of hardy shrubs, all indigenous to the United States. Readily increased by seeds, divisions, and cuttings of the subterranean stems and branches.

Camellia (Tea Plant, Etc.)—Evergreen shrubs and trees from China and Japan. One of the species, *C. theifera*, and its varieties, yields the tea of commerce. The single varieties are propagated by seed, layers, and cuttings of the green twigs planted in sand in frames or under bell glasses. Cuttings of the season's growth, taken off in the autumn and planted in frames in a cool greenhouse, will usually become rooted by the following spring; but green cuttings are usually preferred, as they strike root in a few weeks, if at all. The double varieties are propagated by veneer grafting under glass. The grafted plants should be placed in close frames and frequently watered overhead until the cions have united.

Capparis (Caper Tree).—An extensive genus of evergreen shrubs, of no especial value except the one species, *C. spinosa*, which yields the Caper of commerce. It is a native of Southern Europe, Western Asia, Egypt and North Africa, and is said to be hardy in the Southern counties of England; hence, it will probably thrive in the open air in some of our Southern States. Propagated by cuttings of the ripe wood in sand, under glass.

Caragana (Siberian Pea Tree).—All small, hardy, deciduous shrubs, with one exception. The *C. arborescens* of Siberia grows to a height of fifteen to twenty-five feet in its native country, but ten feet would be considered an extra strong growth in this country. Propagated by seed, layers, cuttings of the roots, and by grafting the low growing species and varieties on the seedling stocks of the *C. arborescens.* If the seeds are kept dry over winter, they will need to be slightly scalded or steeped in tepid water for a few days before sowing in spring.

Carpinus (Hornbeam).—A genus of hardy, deciduous trees, mostly of small size, but with very tough and hard wood. The species are propagated by seeds, which are hard and nut-like, and germinate very irregularly. Some will grow the following spring after sowing, others remaining dormant until the second year. Owing to this uncertainty as to the time of germination, it is always best to sow the seed in the fall as soon as ripe, and in a position where the bed can be freely watered during the following summer, even if but a few plants should appear. But if none are seen by the time warm weather sets in, the entire surface of the bed may be covered to the depth of three or four inches with chaff, moss, hay or some similar material, that will keep the soil moist, thereby saving the labor of applying water. Early the following spring the bed should be uncovered. All kinds of seeds that do not usually germinate until the second season may be safely treated in the same way.

Carya (Hickory, Pecan Nut).—Well-known, valuable, deciduous trees, all natives of the United States. Some of the species, like the Pecan tree and Shell-bark Hickory, yield very delicious and valuable nuts, which are always in demand for home use and export. The cultivation of these noble and valuable trees has, no doubt, been greatly retarded by the prevalence of an erroneous idea in regard to the supposed difficulty or uncertainty attending the transplanting of either young or old trees. This very absurd idea has been repeated and disseminated by men holding high positions, but who could not possibly have had any practical experience in raising or cultivating such trees. These theorists usually recommend the planting of the nut where the trees are intended to remain, which is not necessary any more than it is to plant the seeds of our common fruit trees in such positions. The excuse for this kind of permanent planting is, that the Hickories do not produce a sufficient number of fibers to insure safety in transplanting. If Hickory-nuts are planted in a stiff clay, or any other kind of hard, compact soil, they will send down one or two long, naked tap-roots, but if placed in light, rich sand or loam they will produce a large number of fibrous roots, and may be transplanted with as much certainty of living afterwards as any nut-bearing tree known. I speak from experience, and not hearsay, in this matter. In propagation, select the fresh nuts in autumn, and mix with light soil or sand, and place in heaps in the open ground, or in boxes with good drainage. Early in spring, or as soon as the nuts show signs of sprouting, take them out and drop in rows, placing the nuts two to four inches apart in the row, and the rows four feet apart. Cover the nuts with about an inch of soil. If the seed-bed is light, poor sand, so much the better, but add old, well-rotted stable manure in liberal quantities to the surface, as a mulch. In the fall, or early the following spring, take up the seedlings, either with a spade or with a tree-digger, and if they have long tap-roots, they should be shortened to about one-half their original length, or a little more. If this is done in the fall, and in a cold climate, the plants should be heeled in and well protected from cold, and left in this position until the following spring, when they should be set out in nursery rows, and in heavier soil than recommended for a seed-bed, placing them fifteen to twenty inches apart in the row, and the rows at a distance that will admit of cultivation with plow and cultivator. The trees may remain in the nursery rows until four or five feet high, then removed to the place where they are to remain permanently. While it is true that transplanting usually temporarily checks the elongation of the stem, the number of roots will be greatly increased, and the plants will not only become more bulky, but in a far better condition for making a rapid and vigorous growth in after years. Where there is no danger of the nuts being disturbed by vermin, or of the soil becoming packed and hard during the winter, the nuts may be planted out in rows as soon as gathered in autumn, but my preference is for spring planting in a recently-plowed and freshly-prepared bed for all kinds of nuts and seed. Propagating the Hickories by budding and grafting has never been practised with any very great success anywhere, and when I first published a description of and named the Hale's Paper-shell Hickory-nut, in 1870, I doubt if there was a grafted tree of the Hickory in this country. The making known the existence of this unique and valuable variety, has prompted many nurserymen and other persons to try their skill in grafting these trees, and while their success has not been great, still grafted Hickory trees are no longer unknown, or so very rare. The greatest success, thus far, has been obtained by grafting under glass, using small stocks that have been growing at least one year in pots. For out-door grafting, and in cool climates, terminal grafting is best, either using a splice without a tongue, or by cutting away one side of the cion and thrusting it under the bark at the apex or top of the stock. Wrap with Bass, and cover with grafting clay, and over this a little moss, and then enclose the whole in a hood made of oil paper, leaving it on until the cion shows unmistakable signs of growth. The object of enclosing in oil paper is to prevent the evaporation of moisture, and drying of the clay and cion. The wood or twigs used for cions should be taken from the tree early in winter, and buried in moist sand or packed in sphagnum, and stored in some cool place, where they will remain in a perfectly dormant condition until they are wanted for use. The grafting should not be done until the leaves on the stocks have begun to open. The great difference in the density of the sap of the stock and the cion will insure a rapid flow into the latter. There is another mode of propagating the Hickory, which may be practised when necessary to preserve or increase the number of any choice or rare variety. This is done by exposing the roots to light and air, thereby forcing them to produce buds and sprouts. It is well known that Hickory roots, large or small, when exposed to the light and air, will soon produce buds and sprouts on the exposed surface. This tendency of Hickory roots to produce sprouts from adventitious buds may be taken advantage of in the propagation of valuable natural varieties. The lateral surface roots, at some distance from the main stem—five, ten or twenty feet, according to the age and size of the tree— should be exposed to the light and air early in spring, by removing the soil above them for a space of two feet or more, and leaving them in this condition the entire summer. Sometimes no sprouts will appear the first season, and the exposure will need to be continued during the succeeding one, or until they do appear. Then the main root on which the sprouts have grown may be severed on the side nearest the stem of the tree, and then carefully lifted and followed outward until enough fibers or small roots are secured to ensure the life of the young tree when transplanted. Better formed plants may be secured by allowing the sprouts to remain for a year where they have grown, after severing the root; and if fine, rich soil is thrown in around the base of these sprouts, new fibers will usually appear during the season, and when the plants are removed less of the old or main root will be needed to ensure

Fig. 96.—HICKORY WITH ROOT SPROUTS.

growth after transplanting. In figure 93 a Hickory tree is shown with a number of forced sprouts from the roots in position.

Cassandra (Leather Leaf).—A genus of small native shrubs, closely allied to the *Andromedas*, and of the "Heath Family." Usually propagated by layers or dividing the plants as taken from the bogs and low grounds, where they are to be obtained in abundance.

Cassia (Senna).—A genus of some two hundred species of shrubs and herbs. Very few of the species are in cultivation, or are they of any special interest. All are readily propagated by seed. The ligneous species may be increased by cuttings of the half-ripened shoots under glass.

Castanea (Chestnut).—A genus of a few species, but of many varieties. Propagated by planting the nuts as soon as ripe in the autumn, or preserving them in sand and stored in a cold place until spring, then planting as directed for Hickory-nuts. Varieties are increased by grafting on seedling stocks—splice grafting being the preferable mode, and it should not be done in spring until the sap in the stock is flowing rapidly and the buds have pushed almost into leaf. The cions should be in a dormant state when inserted, and held in place with waxed manilla paper, as waxed cloth does not allow of rapid expansion of the stock when growth begins, and there is danger of strangulation. (For Stocks, see Chapter XVIII.)

Catalpa (Indian Bean).—A genus of handsome, rapid growing trees and shrubs. Cultivated for ornament, and the very durable wood of the larger growing species. Propagated by seed preserved dry during winter and sown in spring lightly covered with fine, rich soil. The Catalpas may also be readily propagated by grafting in spring, and by cuttings of the one year old wood, made in the fall and buried below the reach of frost in the open ground, or they may be preserved in a cool cellar and then planted out in nursery rows early in spring. The Catalpa will not grow from root-cuttings, as has been repeatedly stated in books on forestry and the propagation of plants.

Ceanothus (New Jersey Tea).—An interesting genus of low-growing, pubescent shrubs, with small but pretty flowers. Of the twenty or more species, all are natives of North America. Propagated by cuttings of the ripe wood taken off in the autumn, and by layers. Some of the species may be propagated by root-cuttings placed in a position where they will form buds during the winter months.

Cedrus (Cedar).—A small genus of cone-bearing trees, the Cedar of Lebanon being the best known and most familiar representative. The species are propagated by seed, sown as soon as taken from the very close and compact cones, and in a half-shady position in a cool greenhouse, or in frames in warm climates. The seeds will remain sound for many years if left enclosed in the cones, but soon lose their vitality after removal. Varieties are usually propagated by veneer grafting under glass, late in summer or early spring, using strong, pot-grown seedlings for stocks.

Celastrus (Staff Tree, Bitter-Sweet).—A genus of climbing, deciduous and evergreen shrubs, cultivated for their ornamental foliage and fruit. All the species readily propagated by seed, layers, or cuttings of the mature shoots.

Celtis (Nettle Tree).—A small genus of mostly hardy, deciduous trees and shrubs. The North American species are the largest, but not of any special value. Propagated by seed sown as soon as ripe, and by layers when the trees branch so low down as to admit of this mode of propagation.

Cerasus (Cherry, Laurel, Etc.)—A large genus of shrubs and trees, mostly deciduous; but there are two European and one American species with persistent evergreen leaves. Propagated by seed sown as soon as ripe, or stored in moist sand; also by budding and grafting. (See Prunus and Cherry, under head of Fruit Tree Stocks).

Ceratiola.—A low-growing evergreen shrub of the order *Empetraceæ*, native of South Carolina and Florida. Thrives in dry, sandy soil. Propagated by seed or green cuttings under glass.

Ceratonia (Carob Tree).—An evergreen tree bearing pea-shaped flowers, succeeded by long pods containing a sweet tasted, mucilaginous pulp. Propagated by ripe wood cuttings planted in frames, or from freshly gathered seed.

Cercidiphyllum.—A slender, rapid-growing, hardy, deciduous tree, introduced about twenty years ago from Japan under the name of *C. japonica*. It has small, smooth, heart-shaped leaves, of a purplish color while young. Propagated by green cuttings, made of the tips of the twigs during the summer, planted in sand in a close frame in a propagating house. I have found that the cuttings strike root more readily if slightly dried or wilted before they are placed in the frames. The leaves should be cut away, leaving only two or three of the terminal ones.

Cercis (Judas Tree, Redbud).—A genus of three species, one each in America, Europe, China and Japan; the former two small trees growing twenty to thirty feet high, the latter a stocky shrub six to eight feet. Propagated by seeds kept in moist sand over winter, and by layers.

Chamæcyparis (Cypress, White Cedar, Etc.)—An extensive genus of coniferous trees, according to the recent re-arrangement and classification of the *Coniferæ*. It is represented in this country by three species: the White Cedar (*C. thyoides*), the Lawson Cypress (*C. Lawsoniana*), and Nootka Sound Cypress (*C. Nutkaensis*) of the Pacific Coast. The *Retinispora*, or Japan Arbor-vitæs, are also included in this genus. Propagated by seed, layers, and cuttings of the smaller twigs and

Fig. 97.
CUTTING OF RETINISPORA.

branches taken off in the autumn and planted in frames, or in a cool greenhouse where they will callus slowly; then given a little higher temperature to force out the roots. Some of them produce roots readily and freely, while others under the same treatment will remain fresh and sound for a year before any roots will be emitted. This difference may often be observed in the varieties of a species, as well as in the different species. With both the varieties of a species, as well as species which are found to be difficult to propagate by cuttings taken from plants in the open ground, the most certain mode is to pot a few plants and place them in a greenhouse, and after they have made a new growth of two or three inches take off the young, succulent tips of the branches for cuttings, planting them in sand under bell glasses, or in a close frame, where they can be given bottom heat, applying water overhead daily with a syringe or watering pot, through a fine rose. Cuttings of the species commonly known as *Retinispora plumosa* may be made of the size and form shown in figure 97. Varieties may also be propagated by grafting, using closely allied species for stock. Veneer grafting under glass is the best mode.

Chilopsis (Desert Willow).—A genus represented by one species, indigenous to Texas and westward to Southern California. A slender-growing, small tree, bearing Bignonia-like flowers. Readily propagated by seed, or by ripe wood cuttings, planted in the open ground in warm climates, or under glass in cold ones.

Chionanthus (White Fringe Tree).—Hardy, native, deciduous shrubs, cultivated for their pure white, fringe-like flowers. Propagated by seeds sown in the autumn as soon as ripe; by budding and grafting on the common Ash. (See Fraxin, Chapter XVIII.)

Choisya.—Only one species, and this (native of Mexico) a beautiful shrub, hardy only at the South. Propagated by ripened cuttings, planted in a half-shady position and in rather light soil.

Chrysobalanus (Coco Plum).—Small semi-tropical shrubs bearing edible fruit. Two species indigenous to Florida. Propagated by seed and cuttings of the mature branches, planted in a half-shady position and given plenty of water overhead during dry weather.

Citrus (Orange, Lemon, Shaddock).—Semi-tropical evergreen trees, bearing in their improved state edible fruit. Propagated by seed, budding and grafting in the open air in warm climates, but in cool ones under glass, and usually by veneer grafting. (See Fruit Stocks.)

Cladrastis (Yellow-Wood, Virgilia).—A genus of only two species of large, free-growing, deciduous, ornamental trees. One species is indigenous to Kentucky and Tennessee (*C. tinctoria*), and the other (*C. amurensis*) to the Amoor regions of Asia. Propagated by seeds sown in spring or autumn, and by cuttings of the roots prepared in the autumn and packed in moss or clean sand and stored in a cool cellar until spring.

Clerodendron.—A large genus, mostly tropical and semi-tropical shrubs and vines, bearing bright-colored flowers in long, terminal panicles. Some of the most showy species require a high temperature; others will thrive in an ordinary greenhouse, while one Japanese species is said to be hardy in England. Propagated by seed sown as soon as ripe; by cuttings of the somewhat mature wood in close frames.

Clethra (White Alder, Pepperidge, Etc.)—Ornamental, deciduous shrubs, two or three species indigenous to the United States, and often cultivated in gardens. The foreign species are mostly tender in the North, although a recently introduced Japan species (*C. barbinervis*), is moderately hardy. Propagated by seeds, layers, division of the clumps, and by cuttings of the ripe wood.

Cleyera.—Handsome evergreen shrubs blooming in spring; flowers, white, or yellowish-white; fragrant. One species from Japan and the other from Jamaica. Not hardy in our Northern States. Propagated from green cuttings.

Codiæum (Croton).—A small genus of ornamental-foliaged, evergreen shrubs, natives of tropical countries; consequently require a high temperature to insure health and vigorous growth. There is an immense number of varieties in cultivation, all of which have probably descended from less than a half dozen species. They are more commonly cultivated under the generic name of *Croton*, and are now very popular for decorative purposes. Propagated by seed to produce new varieties, and these by cuttings of the ends of the leading shoots and branches planted in sand, in frames or under bell glasses, giving strong heat and a confined, moist atmosphere.

Colutea (Bladder Senna).—A genus of a few species of ornamental, deciduous, hardy shrubs, with pea-shaped, yellowish flowers, and seeds in bladdery pods; hence, the common English name. Propagated by seeds, or ripe wood cuttings taken off in the autumn and treated as usual with mature wood cuttings.

Comptonia (Sweet Fern).—Only one species, the *C. asplenifolia*, a very common and familiar low-growing shrub, with fragrant, fern-like foliage. It is seldom cultivated. Propagated by dividing the clumps, and by layering in autumn.

Cordia.—A genus of about two hundred species of tropical, evergreen trees and shrubs, two of which are found indigenous or naturalized along the southern border of the United States. A few of the species are cultivated as greenhouse shrubs in European gardens. Propagated by soft or mature cuttings under glass.

Corema (Portugal Crakeberry).—Small, low-growing, Heath-like shrubs. Only two species in the genus. Flowers, diœcious. One species indigenous to our Northeast Coast and Newfoundland, the other to Southwestern Europe. Propagated by cuttings taken off in summer and planted under glass.

Cornus (Dogwood, Cornel).—A genus of about twenty-five species, all but one belonging to the Northern Hemisphere and more than half of these indigenous to the United States. They are mostly hardy, deciduous shrubs and small trees; rarely herbs. A few of the species and their varieties cultivated for ornament and their edible fruit; among the latter, the European Cornel (*C. mas.*) is a familiar shrub in gardens. The American Flowering Dogwood (*C. florida*) is a conspicuous tree in our forest in early spring. A weeping variety (*C. florida pendula*), and

SELECT LISTS OF PLANTS. 267

a red flowering (*C. f. purpurea*), have recently been discovered and are now being extensively propagated by nurserymen by grafting and budding on seedling stocks of the species. Grafting these varieties in the open air is somewhat uncertain, although by using cions composed of two-year-old wood for the splice or wedge, and a short section of one-year-old with a bud above, moderate success may be obtained. But the best mode of grafting is with the veneer graft, under glass, on pot-grown stocks, in August. I much prefer budding in the nursery on stocks that have been headed back in spring, inserting the buds on the new growth of the season, and performing the operation as late in summer as possible, and yet before the stocks have ceased growing. To hasten the development of the buds on the parent tree, pinch off the ends of the young shoots a week or two before the buds are wanted for use. All of the species of the Dogwood are readily propagated by seed, layers, and some—like the "Red Osier Dogwood"—may be readily increased by cuttings of the mature wood.

Corylus (Hazelnut, Filbert).—A small genus of hardy deciduous trees and shrubs, bearing edible nuts. Two small, low-growing species are natives of the United States, but they are seldom cultivated. The European and Asiatic species have yielded an almost innumerable number of varieties; the best of these are extensively cultivated in the Old World, and sparingly so in this country. Propagated by seed, or the nuts preserved in sand over winter, as I have already directed for other kinds of nuts; by suckers, which usually spring up in great abundance about the stems; by layers, and occasionally by budding and grafting. Cuttings of the young shoots of the season, made in the fall and stored in moist sphagnum or sand through winter, will grow quite freely, if planted in a warm, well-drained soil.

Cotoneaster.—Hardy shrubs and small trees, with small white or pinkish flowers, succeeded by ornamental berries late in autumn. There are about fifteen species, a few of the number evergreen in mild winters. Propagated by seed, layers, or grafting on the Quince.

Cratægus (Thorn, Hawthorn).—A genus of some thirty species, mostly hardy, deciduous shrubs and trees; about one-half the number are North American. The double-flowered Hawthorns are popular ornamental trees, propagated by budding and grafting on seedling stocks. To raise seedlings, the fruit should be gathered when ripe, and placed where the pulp surrounding the seed will soften, but not ferment and heat. Then mix pulp and seed with an equal bulk of pure sand, working over the heap until both are thoroughly intermingled. Place all in boxes with good drainage, and on the north side of some building, or where the box will be in the shade, bank up and cover over with soil. The box and seed may remain in this position until the autumn of the next year; then the seed may be taken out and sown in shallow drills, covering them from one to two inches deep, dropping a seed every two inches if sown in single drills, or about the same distance apart if sown in broad drills or beds. As Hawthorn seeds do not usually grow until the second season after sowing, it is always preferable to keep them in a position where water can be given when needed in dry weather, than to sow them fresh and run the risk of having them injured by drouth the following summer. When only a small quantity is to be sown, they may be placed in the seed-bed as soon as the pulp is softened, and the bed kept heavily mulched during the ensuing summer—the mulching removed in the spring of the second season. The main thing to be observed is, to keep them moist and cool until the time arrives for their germination. When the seedlings are one year old, they should be taken up and transplanted into nursery rows, as usually practised with other kinds of stocks.

Croton.—See *Codiæum.*

Cryptomeria (Japan Cedar).—Slender, tall-growing, coniferous evergreen trees, indigenous to Japan. Two species are usually recognized by botanists, but there are several varieties in cultivation. Scarcely hardy in the Northern States, but an occasional specimen survives, when planted in a dry soil, and in a sheltered position. Propagated by seeds and cuttings of half-ripened wood, and planted in sand under glass.

Cunninghamia (Chinese Fir).—Only one species, the *C. Sinensis.* A broad, lance-leaved evergreen coniferous tree, of a very graceful habit, not hardy in the Northern States, but often cultivated as a low bush, and protected in winter. Propagated by seeds and cuttings.

Cupressus (Cypress).—Evergreen trees and shrubs, with small scale-like leaves, mostly compressed and imbricated in four rows on the rather slender branchlets. There are several species, natives of the west coast, but none quite hardy in the Northern Atlantic States. Propagated by seeds and cuttings.

Cydonia (Quince, *Pirus Japonica*).—Well-known, hardy, deciduous trees and shrubs. The common Quince (*C. vulgaris*), is cultivated for its highly-flavored fruit, and the Japan Quince (*C. Japonica*), and its varieties, for their very showy flowers, appearing in early spring. Some of the latter produce very large and spicy-scented fruit. The propagation of the common Quince is usually by cuttings of the mature wood taken off in the autumn, and after the cuttings are made, they are buried in a dry, warm place in the open ground, or in a moderately cool cellar, and planted out in spring. The cuttings may be made from the one-year-old wood, and from this age to that of four or five years old. Layering the branches is also often practised as a mode of propagation; also banking up of the sprouts that appear around the base of old stocks, which have been headed back for the purpose of producing these sprouts. Varieties may be propagated by cuttings, budding and grafting, using inferior sorts or seedlings for stocks. The Japan ornamental varieties are readily increased by cuttings of the young wood of the season, taken after the frost has killed the leaves in the fall, but the most rapid and certain mode is by cuttings of the roots made in the fall, kept in sand or moss over winter, then sown in drills early in spring. The Chinese Quince (*C. Chinensis*), is a very distinct species, bearing fruit of an enormous size, but it is scarcely edible. Propagated by seed, or grafting on stocks of the common Quince.

Cyrilla.—A genus of two species of evergreen trees; one species is found in the Southern States, and the other in the West Indies and South America. Readily propagated by seed and cuttings planted in the open ground in the South, or under glass in the North.

Cytisus (Scotch Broom, Etc.).—A large genus of low, slender-branched shrubs, bearing pea-shaped flowers. All indigenous to the Eastern Hemisphere, and many of the species are known under such local names as Scotch Broom, Irish Broom, Spanish Broom, etc. The hardy species are readily increased by seeds and layers, and the tender, or those cultivated in the greenhouse, by cuttings of the tender shoots planted in close frames, or under a bell glass.

Dacrydium (Tear Tree).—A genus of handsome coniferous evergreen trees, from New Zealand, Tasmania and New Caledonia. All tender, except in the extreme South. Propagated by cuttings of the mature twigs planted in sand under glass, and by seed when it can be obtained in a fresh state.

Daphne (Spurge Laurel, Mezereon).—A highly-prized genus of low-growing evergreen and deciduous shrubs. Some of the species, like the common Mezereon (*D. mezereum*), and the Garland Flower (*D. Cneorum*), have been cultivated in this country for many years; but there are many other species, fully as hardy and valuable, that are rarely or never seen in our gardens. All thrive best in a half-shady position, as the leaves are likely to burn during the hot weather in summer. Propagated by seed, when these can be obtained, by layers, cuttings and grafting; the stronger-growing being used as stocks. *D. Cneorum* is one of the very best of the hardy, low-growing species for cultivation in this country, and it may be increased by layers put down in spring, or by cuttings of the ends of the young shoots obtained from young plants forced in winter, or even from cuttings of the nearly matured wood, if taken off early in the fall and planted in a greenhouse where they will receive only moderate heat—not much above fifty-five or sixty degrees while the callus is forming. Veneer grafting on stocks grown in pots is the best mode, as with other evergreen shrubs.

Darwinia.—Handsome evergreen shrubs, indigenous to Australia. Flowers large and showy, either red or white, produced in terminal fascicles. Cultivated in greenhouses in cold climates. Propagated by cuttings of the young succulent roots placed in a close frame.

Dasylirion (Lily Tree).—A genus of tall-growing diœcious evergreen shrubs, natives of Mexico, and only hardy at the South, but they are fine, showy plants, adapted to greenhouse culture, although rather large unless kept well cut back. Propagated by seeds and cuttings.

Datura (Stramonium).—A genus of shrubs, trees and annuals. A few of the species, like *D. Arborea* (or *Brugmansia candida* of some authors), and *D. meteloides* from California, are cultivated for their large tube-shaped flowers or showy foliage. Propagated by seed; and the shrubby species and varieties by cuttings, which strike root quite readily under glass.

Decumaria.—A handsome climbing shrub, with white, sweet-scented flowers. Indigenous to the Carolinas and Florida. Readily propagated by layers and cuttings. The latter should be made in summer, and planted in the shade, and given plenty of water.

Desfontainea.—A Holly-like, evergreen shrub, native of the mountains of Chili. Said to be hardy in England. Propagated by cuttings, planted either in light loam or sand under glass.

Deutzia.—Well-known, hardy, deciduous shrubs of easy culture. Propagated by cuttings of the mature shoots of the season, made in the autumn and stored in a cool and moist place; planted out in spring. For the slender-growing, dwarf species, like the *D. gracilis*, green cuttings are preferable, taken from plants forced under glass in winter. These green cuttings should be taken off with a heel, or close to the old stem, and then set in close frames; they will strike root in a very few days.

Diervilla (Weigela).—A genus of hardy, deciduous shrubs, with showy, funnel-shaped flowers, varying in color from pure white and yellow to deep rosy purple. The Chinese and Japanese species and their varieties are far more common in cultivation than our two indigenous species, viz., *D. trifida* and *D. sessiliflora*. Propagated very readily by either ripe wood cuttings in the open air, or green wood under glass.

Dimorphanthus.—See *Aralia*.

Elæagnus (Oleaster, Wild Olive).—A small genus of evergreen and deciduous shrubs. One species, the *E. argentea*, is a native, and sometimes cultivated; also one or two species from Europe and Japan. Propagated by seed, layers and root-cuttings.

Elliottia.—A genus of only three species, one indigenous to our Southern States, and two are found in Japan. The first is a deciduous shrub, growing four to ten feet high, bearing small, white flowers in terminal racemes, resembling those of the *Andromeda*. Propagated by soft wood cuttings under glass.

Epacris.—Small evergreen shrubs and trees, mostly from the Indian Archipelago, Australia and Polynesia. All are tender, and require the protection of a warm greenhouse in winter, but may be planted out in summer. They are highly valued for the great number of flowers which appear during the winter months. Readily propagated by cuttings of the tips of the growing shoots, taken off during the winter, and planted in a close frame with bottom heat.

Erica (Heath).—An extensive genus of nearly, or quite, four hundred species of evergreen shrubs, with small and slender branches. A large majority of the species are native of the Cape of Good Hope, and are tender in cool climates. There are, however, a few species native of Europe, and hardy, but they require a moist soil and a rather shady position. The Heaths are very popular plants, and extensively cultivated in European gardens and greenhouses, but seldom seen in any large numbers in this country, as they require special care and attention in order to produce fine, healthy specimen plants. Propagated by seed for producing new varieties, and by cuttings of the ends of the mature twigs, also from the young growth of what are called the soft-wooded species. The cuttings should be short—not more than two inches long, and one-half this length will answer equally as well. Very clean sand should be used in which to set the cuttings of these plants, and the pots or boxes given good drainage.

Erythrina (Coral Tree).—A genus of about thirty species of trees and shrubs, bearing showy pea-shaped flowers. They are principally natives of tropical countries. One species (*E. herbacea*), with slightly woody stem, indigenous to our Southern States. The flowers of all the species are coral-red, scarlet or copper colored, hence the common name of "Coral tree." The Cockscomb Coral tree, from Brazil (*C. Crista-galli*), is the best known and most common species in cultivation in this country. The plants are usually stored in a warm cellar or cool greenhouse during winter, and planted out in summer. Propagated by cuttings of the young shoots as they start in spring, taken off with a heel, and planted where they will receive a little bottom heat.

Eucalyptus (Australian Gum Tree, Fever Tree).—An immense genus of very large, broad-leaved evergreen trees, mostly natives of Australia. Specimen trees of some of the species are said to have been found in the forests of Australia that are over four hundred feet high, with stems more than fifty feet in circumference. The Blue Gum tree (*E. globulus*), has been planted quite extensively in California as a timber tree, and also in the malarious districts of Italy, where it is said to have a beneficial effect in checking malarial fevers. Propagated by seeds imported from Australia, and which usually grow very readily. The seed should be sown in shallow boxes, and the seedlings transplanted when only a few inches high.

Euonymus (Burning Bush, Spindle tree).—A genus of ornamental, deciduous and evergreen shrubs and small trees. The flowers are mostly small and inconspicuous, but in several of the species they are succeeded by very brilliant-colored berries remaining on the plants until late in autumn. Of the evergreen Japanese species (*E. Japonicus, E. radicans, etc.*), there are several variegated-leaved varieties in cultivation, some of them requiring protection during the winter in our Northern

States, while others are quite hardy. The deciduous species are propagated by seed and ripe wood cuttings, and the evergreen varieties by cuttings of the young wood under glass. They may also be propagated by layers, but cuttings produce the best-formed plants. The large fruited and broad-leaved Euonymus (*E. latifolius*), is usually propagated by budding and grafting on stocks of the larger-growing species, and the low, trailing species may also be grafted on the same kind of stocks, the evergreen thriving on the deciduous.

Euphorbia (Milkwort, Poinsettia, Etc.)—An immense genus of several hundred of species of trees, shrubs and herbs, the greater part abounding in a milk-like juice, often containing on acrid and poisonous principle, others containing valuable medicinal properties. The flowers are small, unisexual and crowded in numbers at the ends of the main shoots. A few shrubby species are cultivated in greenhouses, usually under the name of *Poinsettias*, the bright-colored bracts surrounding the umbel of flowers being used for decorative purposes. The winter-blooming, tropical species require a high temperature to insure full development of the showy, leaf-like bracts. Propagated by cuttings, and they strike root freely in a temperature of about seventy-five to ninety degrees.

Eurya.—A small genus of evergreen shrubs from China and Japan. A variegated-leaved variety of *latifolia* is occasionally cultivated as a half-hardy shrub in our Northern States. Propagated by green cuttings under glass.

Exochorda (Large-flowered Spiræa).—A very large, hardy deciduous shrub, introduced about twenty-five years ago from China, under the name of *Spiræa grandiflora*, but, owing to its being difficult to propagate, it has not as yet become common. Green cuttings taken from plants forced in a greenhouse strike root more or less freely, but the usual practice is to splice graft cions of the ripe wood on pieces of the roots, then plant in hot-beds or in boxes in a moderately warm greenhouse. It is also reported that in France ripe wood cuttings planted early in fall in the open ground strike root quite readily.

Fabiana.—A neat little shrub with white flowers, and a well-known, common greenhouse plant. Propagated by seeds and cuttings.

Fagus (Beech).—A genus of large and valuable timber and ornamental trees. All are deciduous except two or three species found in South America, but of which little is known. The triangular-shaped nuts are well known among the people of both Europe and America. Propagated by seed, which should be sown as soon as it is taken from the tree, or stored in moist sand and in a cool place through winter. Varieties are propagated by layers, and by budding and grafting in the open air. In grafting, the cions should be set low down, and as near the roots as convenient for working handily.

Faramea.—A neat little evergreen shrub from the West Indies. Cultivated in greenhouses for its white, fragrant flowers, which are about the size of those of the Cape Jasmine. This species (*F. odoratissima*), is known by various names, such as *Coffea occidentales, Ixora Americana,* etc.

Ficus (Fig Tree).—A very large and important genus of evergreen shrubs and trees. One species, the *F. Carica,* yields the cultivated figs so well known to all civilized and to many barbarous nations. Another species, the *F. elastica,* yields a part of the India-rubber of commerce, while another, *F. Indica,* is the celebrated Banyan tree of India. The varieties of the common Fig and India-rubber tree are readily propagated by cuttings of the green wood or mature shoots planted in rather coarse sand or pulverized brick, in the open ground in warm climates, and under glass in cool ones.

Fitzroya.—A coniferous, evergreen tree from Patagonia, with the habit of the Weeping Cypress. Only one species in cultivation, the *F. Patagonica,* and this is far from being common. Propagated by seeds and cuttings of the half-matured twigs and leading shoots under glass.

Fontanesia.—Large and handsome ornamental shrubs, closely related to the common Olive tree. One species, from Syria (*F. phillyræoides*), has long been cultivated in our gardens, but the Chinese species (*F. Fortunei*) is not so well known. Propapated by layers and cuttings of the mature wood under glass.

Forsythia (Golden Bell).—A small genus of handsome, early-blooming, hardy shrubs from China and Japan. Three species are recognized in the late botanical works, viz., *F. viridissima,* from China in general; *F. Fortunei,* from Pekin, China, and *F. suspensa,* from Japan. The latter is very distinct from the former two, having long, slender branches, drooping or trailing on the ground. The leaves are quite small, ovate, and on some branches all will be trifoliate, somewhat after the form of a clover leaf. The flowers are usually of a bright yellow, appearing several days earlier than on the first-named two; still, with this wide difference in habit of growth, size and form of leaves, and time of blooming, I am quite certain that it is only a garden variety (or it may be a wild one), of the first, or *F. viridissima.* My reason for making this statement is that from a large number of seedlings raised from the seed of an isolated plant of *F. suspensa,* at least nine out of every ten have assumed the upright habit, with the strong, stout canes and large, entire leaves of the two first-named species. In these seedlings we have an instance of reversion to the original type of the most pronounced kind; for there could not possibly have been any crossing or hybridizing with any other closely allied plant. Many of the seedlings of this Weeping Golden Bell are now large, old plants, so well established that their varietal characters may be considered as fully developed. All the varieties are readily propagated by cuttings of the ripe wood, made in either fall or spring, and planted in nursery rows.

Fothergilla.—A genus of one, or at most two, species of low-growing, deciduous shrubs, native of the swamps and low grounds of the Southern States. *F. alnifolia* is sometimes cultivated for its rather showy, white flowers, appearing in spring. Propagated by seed and layers.

Fraxinus (Ash).—A genus of many species, mostly large, deciduous trees of cold climates. Highly valued for their timber, as well as for ornamental purposes. There are also a large number of varieties in cultivation. The species are usually propagated by seed, which should be gathered when ripe, in autumn, and either sown immediately, or packed in moist sand and stored in a cool place, and then sown in spring, covering lightly with rich, loose soil. Sometimes the seeds will remain dormant and not germinate until the following spring, and when this is likely to occur, the seed-bed should not be permitted to become dry, but given water or covered with some light, chaffy material to aid in retaining moisture. The seed should also be examined when gathered, to see if they contain a good, plump kernel, as seeds on isolated trees are often defective or false, owing to non-fertilization of the flowers. Varieties are propagated by budding and grafting in the open air. (See Chapter XVIII., on Tree and Shrub Stocks.)

Fremontia.—An interesting, large, ornamental, deciduous shrub, with yellow flowers. This genus is closely allied to the Basswoods or Lindens. One species, *F. Californica*, is known locally in California as "Slippery Elm," the inner bark being used as a substitute for that of the *Ulmus fulva*. *Vide* Prof. Rothrock, botanist to Wheeler's expedition. Propagated by seed and layers.

Fuchsia (Ladies' Eardrops).—Well known and popular, tender, ornamental shrubs, mostly native of Mexico and South America. The *F. arborescens* of Mexico is said to grow to a height of ten to twelve feet in its native habitats, but there are few other species and varieties that exceed five or six feet. All are readily propagated by seeds and by cuttings of the young, tender shoots.

Gardenia (Cape Jessamine).—A very popular genus of evergreen shrubs, producing sweet-scented flowers. The double flowered variety of *G. florida*, from China, has long been a favorite greenhouse plant in our Northern States, and a common garden shrub in the Southern. In the South it is readily propagated by layers, but in cool climates by cuttings of the young, tender shoots, planted in close frames in the propagating house, or in an ordinary warm greenhouse.

Garrya.—A small genus of evergreen shrubs, of the order *Cornaceæ* (Dogwoods), native of the Rocky Mountain regions and the Pacific Coast. The young branches and branchlets are somewhat four-angled; fruit, blue or purple. Seldom cultivated, and little is known of their propagation further than by seed and layers.

Genista.—See *Cytisus*.

Gleditschia (Honey Locust, Etc.)—A genus of deciduous trees with elegant, pinnate leaves, all of graceful habit; but most of the species produce simple or branching thorns of great strength as well as length, and for this reason they are objectionable, and rarely planted except for hedges. There are, however, several unarmed varieties of the most hardy and valuable species, and these are well worthy of general cultivation as ornamental trees. The species are propagated by seed, sown as soon as gathered, and removed from the pods, or mixed with moist sand and buried in the open ground. If the seeds are kept dry over winter they should be scalded or steeped in warm water for a day or two before sowing. It is always advisable to soak old seeds and keep them in a warm place until the sprouts begin to appear, before sowing; for, if they do not germinate while in a warm place in the house, they will not grow when sown in the field. The varieties may be propagated by grafting on seedling stocks in the open air. (See Chapter XVIII., Selecting Stocks.)

Gymnocladus (Kentucky Coffee Tree).—Only one species, the *G. Canadensis*, common in the Middle and Western States, thriving best in rich, moist soils. A large, deciduous, ornamental and timber tree, with very long, bi-pinnate leaves. Seeds produced in long, broad pods—large and very hard when mature, requiring the same or similar treatment as those of the Honey-locust. English nurserymen are said to propagate the Kentucky Coffee Tree by cuttings of the roots, but there is no occasion to resort to this mode here, as seeds are to be obtained in abundance.

Halesia (Snowdrop, or Silver-Bell Tree).—A small genus of very ornamental, deciduous shrubs, of the Styrax Family. Only three species, all indigenous to the Southern States; one extending as far north as Virginia, but all hardy in much colder latitudes, as I have never known them to be injured in my grounds in Northern New Jersey. Propagated by seeds, layers, and cuttings of the roots. Seeds, a hard, bony nut, enclosed in a persistent, fibrous husk. The seeds do not usually germinate until the second year, and should be sown where they can be kept moist during the first summer. By covering the seed bed with mulch, this condition can be secured with but little trouble or expense. Seeds self sown, under Pine trees, in my grounds, germinate readily without any artificial aid. (See Chapter XVIII., on Selecting Stocks, Styrax, etc.)

Halimodendron.—A handsome little deciduous shrub from Siberia, closely allied to the *Caragana*, which see for propagation; also Chapter XVIII., Stock for *Halimodendron*.

Hamamelis (Witch-Hazel).—Tall, hardy shrubs, of no especial value or beauty, having the rather singular habit of blooming late in fall or early winter, and ripening the seeds the following autumn. Flowers small, bright yellow, in clusters in the axils of the leaves. A species has recently been introduced from Japan (*H. Japonica*), also one variety. Propagated by seeds and layers. The Japan species may be grafted on the American, but it is rather difficult to make the cions unite in the open air.

Helianthemum (Rock Rose).—An immense genus of shrubby and herbaceous plants. The shrubby species often cultivated in conservatories for their large and showy flowers—usually under the generic name of *Cistus*. The shrubby kinds are propagated by seed, and green cuttings under glass.

Hibiscus (Rose of Sharon, Rose-Mallow).—A genus of about one hundred and fifty species, widely distributed around the world. Mostly stout, herbaceous plants, with large, showy flowers; a few shrubby species, like the common garden Althæa or Rose of Sharon (*H. Syriacus*), and the Rose of China (*H. Rosa Sinensis*), largely cultivated as greenhouse plants in cold climates. Of the two shrubby species named, there are a large number of varieties, all highly prized for their showy flowers. Propagated by seeds, and cuttings of either the green wood under glass, or ripe wood in the open ground. Ripe wood cuttings should be made in the fall, in cold climates, and stored where they will not freeze during the winter. They should also be kept rather dry, too much moisture being very injurious. The same is true of the plants raised from ripe wood cuttings; and they should be dug up in the autumn of the first season and heeled-in, either in a cool cellar or some dry place in the garden.

Hovenia.—A large shrub bearing edible fruit, from Japan and Nepal. It is said that the fruit tastes somewhat like a good Pear. The Japan species (*H. dulcis*), is reported to have produced fruit in the open ground in Philadelphia, but in my grounds the leaves and stems are invariably killed by the first hard frost in autumn. Propagated by seeds, and by root-cuttings.

Hydrangea.—A genus of elegant drawf, mostly deciduous, ornamental shrubs. There are three shrubby species native of the United States, and about a half-dozen in China and Japan. There are, however, many garden varieties of the various oriental species. Propagated by suckers, layers and cuttings of the green and half-ripened wood. Of some of the hard wooded species, like the Oak-leaved (*H. quercifolia*), and the recently introduced Japan Hydrangea (*H. paniculata grandiflora*), green cuttings are most certain, if taken from plants forced under glass.

Hypericum (St. John's-wort).—There are more than one hundred and fifty species in this genus, mostly herbaceous plants of little value or beauty. Of the shrubby species, a few are cultivated in gardens, among these Kalm's St. John's-wort (*H. Kalmianum*), is probably the most common. The hardy species are propagated by seeds and division of the clumps, and the tender ones by green cuttings under glass.

Idesia (Japan Cherry).—A handsome, rapid-growing fruit and ornamental tree from Japan. The one species introduced—*I. polycarpa*—has not proved to be hardy as far north as New York City, but thrives in the South. Readily propagated by root-cuttings made in the fall and planted out in the spring.

Ilex (Holly).—Mostly evergreen shrubs and small trees, with prickly leaves, and small white or yellowish flowers in axillary clusters, succeeded by small berry-like fruit. Species of this genus may be sought in botanical works under such generic names as *Berberis*, *Mahonia*, *Prinos* and *Myginda*. The European Holly (*I. aquifolium*), and its many varieties, is not hardy in our Northern States, but the American Holly (*I. opaca*), thrives in sheltered positions, as far north as the southern counties of the State of Massachusetts. Several trees in my grounds have withstood the cold and produced a fine crop of berries during the past twenty years. The European and true American Holly are increased by seed, which should be gathered late in autumn, or during the winter, and placed in a vessel that will hold water and kept wet for a few days, then the pulp washed off clean. The seed should then be mixed with moist sand, and exposed to the cold in the

Fig. 98.—CION OF AMERICAN HOLLY.

open air, or sown in a bed and covered about an inch deep, and the whole surface of the seed-bed well mulched. The seed does not usually sprout until the second year, and if allowed to remain very dry they will seldom germinate at all. Varieties are propagated by veneer graftings under glass in August. The American Holly is highly prized for its bright, red berries, which remain upon the plants all winter, but as it is not every tree that produces berries in abundance, grafting has recently been resorted to for the purpose of securing a stock of productive plants for cultivating in pots, as well as in the open ground. The cions for this purpose should be taken from productive native trees, the lateral berry-bearing twigs, as shown in figure 98, being selected for this purpose. I have in my grounds one tree of the American Holly, not less than thirty years old, that has never borne a berry, although it blooms freely, but another of the same age, near by, is loaded every season with its bright scarlet fruit.

Illicium (Anise Tree).—A small genus of evergreen Anise-scented shrubs of the Magnolia family, two of which are native of Florida, one of China and another of Japan. Propagated by seeds and cuttings of the ripened shoots planted under glass in warm climates. All thrive best in a moist or wet soil.

Indigofera (Indigo Plant).—A large genus of annual, perennial, herbaceous, and shrubby plants. A few of the shrubby kind are cultivated in greenhouses, as they are very ornamental when loaded with their red or purplish flowers. Readily propagated by cuttings of the young, tender shoots.

Itea (Willow Shrub).—A neat little native shrub (*I. Virginica*), bearing long, slender racemes of minute, fragrant flowers. Usually found in low grounds, from New Jersey southward, but thrives in any good soil, and is quite hardy. Propagated by layers and suckers.

Jasminum (Jessamine).—Evergreen and deciduous climbing shrubs from the Old World, cultivated under glass in cold climates, but many of the species succeed in the open ground in the South. Flowers fragrant, and mostly white or yellow. Propagated by layers and cuttings of either the ripened or green wood.

Juglans (Walnut, Butternut).—A genus of long-known, nut-bearing trees. The so-called English Walnut (*J. regia*), is a native of Asia, but has been cultivated for many centuries in the warmer countries of Europe, and a large number of varieties produced. The Black Walnut (*J. nigra*), is one of our well-known timber trees, bearing large, round nuts of a strong, rank flavor. The Butternut (*J. cinerea*), is also a valuable timber tree, the nuts oblong, with rough ridges; kernel sweet, pleasant tasted, but very oily. A closely allied species to the last (*J. Californica*), is a native of California and Arizona, and one other species (*J. rupestris*), is found from Arizona to Texas. Propagated by seed which should be planted as soon as ripe, or stored in a cool, moist place during winter. If planted in light soils, the seedlings produce a large number of fibrous roots, and are readily transplanted without loss, but when grown in rather firm soils, the seedlings will produce long, naked tap-roots, with few fibers. Varieties are propagated by grafting by ordinary modes in the open ground in mild climates, but in cool ones, out-door grafting is uncertain and seldom successful. Budding is preferable to grafting, for if the bud fails, the stock is not seriously injured. The buds should be plump, rather mature, and then inserted into a rapid-growing stock or branch, through which the sap is flowing rapidly at the time of performing the operation. The English Walnut is not quite so difficult to propagate by budding and grafting as the Hickory, still it requires care and skill to insure success, either in the open air or under glass.

Juniperus (Juniper, Red Cedar).—A large genus of coniferous evergreen trees and low shrubs, native of the Northern Hemisphere. Wood fine grained, not resinous. The heart wood usually of a reddish color, and fragrant, and exceedingly durable. Propagated by seeds and cuttings. The seeds are very hard and bony, and unless the shell is softened by some chemical application they seldom germinate until the second season, even when exposed to frost and kept constantly moist. The usual method of treating the seeds is to gather them in the fall when fully ripe, and either mix with strong, moist wood ashes, or pour some strong potash water over them, leaving them to soak and soften for two or three days; then rub the berries until the outer coat is removed. A little sharp sand added will assist greatly in cleaning the seed. Wash out the sand and other foreign matter by placing the seed in a sieve and pouring the water over them. Sow immediately in a bed in the open air, and cover the seed about one-half inch deep. Over the surface scatter leaves, chaff, or some similar light material. In spring remove the mulch, and if the plants appear, protect them from the direct rays of the sun, but if they fail to come up, cover the bed again with mulch, and leave it undisturbed until the following spring. Most of the Junipers may be propagated by cuttings of the young shoots planted in sand under glass, or of the mature wood taken off in the fall and set in cold frames, where they will receive only slight protection during the winter.

Kalmia (Laurel, Calico Bush, Spoonwood).—A genus of North American evergreen shrubs; only one species, the *K. latifolia*, growing to the height of twenty feet and over. All very ornamental, and highly prized in Europe, but only sparingly cultivated in this country. Propagated by seed and by layers abroad; but wild plants from the woods and fields can be obtained to supply the demand in this country.

Kerria (Golden Corchorus).—One species in cultivation (*K. Japonica*), but of this there are several varieties; one with double yellow flowers is very common in gardens; the single flowered is more rare, but really the most desirable of the two. There is also a variegated-leaved variety. Easily propagated by suckers or cuttings of the mature shoots, planted in the open ground in fall or spring.

Kolreuteria (Bladder-pod).—A genus of one species, viz., *K. paniculata;* a small tree somewhat resembling the common Sumac, having pinnate leaves of numerous leaflets. Flowers small, yellow, in large terminal panicles, succeeded by large bladdery pods, containing large, shot-like, black seed. Propagated by seeds, layers and cuttings of the roots.

Laburnum (Golden Chain).—The common Laburnum (*L. vulgare*), of Europe, is placed in the genus *Cytisus* by some botanists, while others have separated it because of the difference in the general appearance of the plants, and the absence of the *Caruncle*, which is present on those of the true *Cytisus.* Propagated by seeds, layers, suckers, grafting and budding. (See Chapter XVIII., Stocks).

Lagerstrœmia (Crape Myrtle).—A splendid genus of deciduous shrubs from China and the East Indies. Flowers with wavy, crisped petals in large panicles, and of a light red or white color. All popular shrubs, cultivated under glass in cool climates, and in the open air in warm ones. Propagated by layers and by cuttings of the young, tender shoots, placed in a confined, rather moist, and warm atmosphere.

Lantana.—Low-growing, semi-tropical shrubs, producing pink, yellow and orange colored flowers in great profusion. Cultivated extensively for bedding out in summer. The plants grow rapidly, and are very showy when planted in masses. Propagated by cuttings in sand under glass.

Larix (Larch, Tamarack).—Deciduous, coniferous trees, thriving best in swamps, or cold, moist climates. Valuable timber trees. Native of North America, Europe and Japan. Species propagated by seeds, which should be kept dry over winter, and sown early in spring in a finely prepared seed-bed, the seed to be but lightly covered with sand, or light, fine leaf-mold. The young plants should be shaded, either with lath screens, or with branches of trees, until they are two or three months old, as they are liable to be burned off by the hot, scorching rays of the sun in our climate. The seedlings may remain in the seed-bed until two years old, and then transplanted into nursery rows early in the spring. Varieties and rare species are propagated by veneer grafting under glass in August, or by cleft and splice grafting early in spring in the open air. The former mode is preferable, and the most certain. (See Stocks, Chapter XVIII.)

Laurus (Laurel, Bay Tree).—The natural order *Lauraceæ* contains about fifty genera, and an immense number of species. The Cinnamon, Camphor tree, Sassafras, Spice Bush, Sweet and Red Bay, are all familiar plants of this great family. In Europe the Bay tree (*L. nobilis*) is the shrub usually referred to under the name of Laurel, while in this country the shrubs called Laurels do not belong to the Laurel Family. The Laurel or Bay tree of Southern Europe, and its varieties, is sometimes cultivated here in conservatories for their highly perfumed foliage. Propagated by cuttings of the half-ripened wood.

Ligustrum (Prim, Privet).—Ornamental evergreen and deciduous shrubs from Europe and Asia. The common European Privet (*L. vulgare*), is quite hardy in the Northern States, where it is often planted for ornamental hedges and screens. Some of the Chinese and Japanese species are hardy, if given a slight protection during the winter months. All the species are readily propagated by cuttings planted in the open ground, or under glass.

Limonia.—Evergreen trees and shrubs of the Citrus Family, requiring the same treatment and culture as mentioned for Citrus.

Lippia (Lemon-Scented Verbena).—The common species cultivated in greenhouses, from Chili, is better known under the name of *Aloysia citriodora*, but some of our botanists consider *Lippia* as the proper name of the genus. There are several other species, but they are rarely cultivated.

Liquidambar (Sweet Gum Tree).—A genus of one North American and one oriental species. The first is a widely-distributed, handsome forest tree of large size, with bright green, star-shaped leaves which usually change to a dark red or crimson color in autumn. Propagated by seed sown as soon as ripe in the fall, in very moist soil. Some of the seed may germinate the following spring, but they usually remain dormant until the second season, and for this reason it is well to sow them where the bed can be watered during dry weather in summer.

Liriodendron (Tulip Tree, White Wood).—A very large and widely-distributed indigenous forest tree, valuable for its timber, and highly prized as an ornamental tree. Only one species, the *L. Tulipifera*. Propagated by seeds sown as soon as ripe in the autumn, and covered about a half-inch in depth with leaf-mold or other light soil. The seedlings are inclined to make very long, slender tap-roots, and they should be frequently transplanted while young, if they are to be moved when large or several feet in height. The transplanting should always be done in the spring, for the Tulip tree and other members of the Magnolia Family have rather soft, spongy roots, liable to injury from cold and moisture, if disturbed in the fall, and, while seedlings and larger trees may be taken up and heeled-in and given protection during the winter months, it is seldom safe to plant them out in nursery rows or elsewhere in the fall—at least not in cold countries.

Lonicera (Honeysuckle.)—A very extensive genus of ornamental, upright and climbing shrubs. In some botanical works, the species of this genus are separated into two groups, the upright-growing under the above name, and the climbing under that of *Caprifolium*. All, however, belong to the *Caprifoliaceæ*, or Honeysuckle Family. All readily propagated by layers put down in fall, or early spring, also by ripe wood cuttings in the open ground, and green cuttings under glass.

Lycium (Matrimony Vine).—Numerous species; mostly hardy, but a few from the Cape of Good Hope are tender and cultivated in greenhouses. The best known is the common Matrimony Vine (*L. vulgare*), from Southern Europe, a slightly thorny, half-climbing shrub, with small, greenish-purple flowers, succeeded by red berries. Easily propagated by cuttings of the mature one-year-old wood, or by seed.

Lyonia.—See *Andromeda, Cassandra, Oxydendrum*.

Maclura (Osage Orange, Bow-Wood).—A well-known native tree of our Southwestern States. Formerly extensively planted for farm hedges, the young branches being well furnished with strong, sharp spines. The best mode of propagation is by seed sown in spring, in light, rich soil. To hasten germination the seed should be soaked in warm water for two or three days before sowing.

Magnolia (Cucumber Tree, Etc.)—A genus of highly ornamental deciduous and evergreen trees and shrubs, natives of America, China and Japan. A few of the species grow to a large size, and are valuable timber trees, notably the common Cucumber tree (*M. acuminata*). Propagated by seeds, layers, budding and grafting. The seed should not be allowed to get thoroughly dry, but as soon as removed from the pulpy covering, be mixed with sand and sown immediately, or buried in boxes in the open ground for the winter; then taken out in early spring and sown in frames where water can be supplied, and the young plants shaded when they first appear above ground. Layers put down in early spring and notched, or a tongue made on the under side, will usually become well furnished with roots the first season; if not, they should be allowed to remain undisturbed a year longer. Budding is a rather uncertain mode of propagation, but with rapid growing stocks and plump mature buds moderate success may be obtained, even in cool climates. Grafting under glass, employing stocks grown in pots, is the most certain mode for increasing varieties and rare species. (See Veneer Grafting, Chapter XVII, and for Stocks, Chapter XVIII.)

Mahonia.—See *Ilex, Berberis, Aquifolium.*

Malvaviscus (Scarlet Mallow).—Evergreen shrubs of the "Mallow Family," from Texas and tropical America. Flowers scarlet, of a peculiar convolute or twisted appearance, not opening broad, as in the Abutilon and other closely allied plants. The most familiar species is the *M. arboreus*, often cultivated under the name of *Achania Malvaviscus.* It does not seed freely in cultivation, but is easily propagated by cuttings made of the short side shoots, removed with a heel, or close to the old wood.

Mangifera (East India Mango).—Evergreen tropical trees, bearing very large fruit, that of some varieties of delicious flavor; others have the taste and fragrance of turpentine. The Mango is called the "Apple of the tropics," and it is now largely cultivated in the West Indies and throughout tropical America. It is occasionally cultivated in conservatories. Readily propagated by cuttings of the ripened wood planted in sand.

Melastoma.—A genus of tropical evergreen shrubs, mostly with large purple flowers, blooming freely in summer. The petals soon drop when cut from the plant; only a few of the species cultivated in this country. Readily propagated by cuttings under glass.

Melia (Pride of India, China Tree).—A genus of large, handsome, ornamental trees, mostly evergreen and native of tropical countries. One species, the *M. Azedarach*, a deciduous tree from Persia, has long been cultivated in the Southern States under the name of China Tree. The fruit resembles in size and form the common cherry, and it is eaten by birds. Propagated by seeds, which should be sown in the fall, or as soon as ripe.

Mespilus (Medlar).—Low-growing, hardy trees closely allied to the common Pear and Quince, and of the Rose Family. Propagated by seed, and grafting on various stocks. (See Chapter XVIII.)

Mezereum.—See *Daphne.*

Morus (Mulberry).—A genus of few species, but these extending around the world in the Northern Hemisphere. They are principally trees of moderate size, but of great importance to mankind. The leaves of the oriental species supply the silkworm with food, the timber is also valuable for fuel and other purposes, and some of the varieties, like the Downing Mulberry, produce excellently flavored fruit. The species are propagated by seed, layers, and by cuttings of the mature wood taken off in the fall. Most of the cultivated varieties are readily increased by cuttings, but an occasional one is found rather difficult to propagate in this way, and root-grafting in winter or spring is resorted to in its propagation.

Myrtus (Myrtle).—A genus of evergreen shrubs and trees, very few cultivated outside of tropical countries. The common Myrtle (*M. communis*), and its varieties, are well known greenhouse shrubs. Propagated by cuttings of the green shoots under glass.

Negundo (See *Acer*, Maple).

Nerium (Oleander).—A genus of showy evergreen shrubs, natives of the East Indies and Southern Europe. There are but few species, but of the oldest and best known there are many varieties in cultivation. They are very popular greenhouse plants, blooming the greater part of the year. Readily propagated by cuttings planted in sand, and then kept moist and warm. The half-ripened shoots will produce roots, if the lower ends are kept immersed in water alone.

Nyssa (Tupelo, Sour Gum Tree).—A genus of North American deciduous trees, usually growing in moist soils or near the borders of streams. Flowers small, greenish. Fruit, a one-seeded drupe, and in some of the species edible. The Tupelos, although greatly admired for their deep, glossy green foliage, which assumes a bright crimson color in autumn, are rarely seen in cultivation, owing, it is said, to the difficulty of making transplanted trees live. To prepare the trees for safe removal, when of good size, the seedlings should be transplanted every two or three years, and the soil above their roots covered with mulch. Propagated by seed, which seldom germinates until the second year; and the seed-bed must not be permitted to become dry from the time of sowing the seed, until the plants have appeared and become well established. A safe and certain mode of obtaining good plants for transplanting is to raise them in pots plunged in a frame in the open ground.

Olea (Olive).—The common Olive tree (*O. Europœa*) is the most important species of the genus, yielding the well-known, edible, oily fruit of commerce. There are, however, quite a number of species of trees and shrubs belonging to this genus, that are well worthy of cultivation for ornamental purposes, but not for their fruit. The Olive is readily propagated by cuttings of the ripe wood, old branches taking root even more freely than the one-year-old. In warm climates, where the Olive flourishes, the cuttings are planted in the open ground in the autumn. In European countries large truncheons or cuttings are used instead of those of moderate size and length, but for no better reason than because it is the general practice or custom, just as long cuttings are used in propagating the grape in the same countries, instead of very short ones, as in this country. Chips cut from an old Olive tree stem will readily produce sprouts, if planted in a warm soil and kept moist; in fact, the entire surface of this tree will produce adventitious buds very freely, if placed in a position to receive heat and moisture.

Osmanthus (Japan Holly).—A genus of neat little evergreen shrubs, of the "Olive Family," from Japan. Flowers small, white and very fragrant. Not hardy in the north, but often cultivated in greenhouses. There are several species or varieties in cultivation under such names as *O. aquifolium*, *O. fragrans*, and *O. ilicifolius*. Propagated by cuttings under glass.

Ostrya (Iron-Wood, Hop-Hornbeam).—Slender, deciduous, hardy trees, with very firm, hard wood. One American species, *O. Virginica*, and one European, *O. vulgaris*. Propagated by seed, layers, and grafting in the open air.

Oxydendrum (Sorrel Tree).—Only one species, the *O. arboreum*, native of Ohio and south. A rather scarce tree, growing fifteen to twenty, and even fifty feet high; leaves rather acid to the taste when young; very smooth, glossy green when mature, but in autumn changing to a dark crimson color, even before they are touched by frost. Propagated by seed sown in frames, lightly covered with fine soil, and kept shaded and moist until they germinate. The young plants are exceedingly delicate, and require great care and attention to keep them growing through the first season. Layering is practicable, if strong, old plants are headed back for the purpose of producing sprouts suitable for layers, but roots are produced very slowly from layered shoots, however carefully the operation is performed.

Pæonia (Pæony).—A genus of well-known, tuberous-rooted, herbaceous and shrubby plants, with very large, showy flowers. Mostly natives of Siberia and China. One herbaceous species (*P. Brownii*) indigenous to California and northward. Herbaceous varieties propagated by dividing the crowns bearing a terminal bud on each division or tuber, as adventitious buds are seldom or never produced below the crown. Shrubby species are propagated by division, layers, and cuttings taken off late in summer, with a heel or a piece of the preceding year's wood attached, and planted in a cool greenhouse, or in a frame where they can be given sufficient protection to prevent freezing during the winter.

Grafting is also practicable, using the large fleshy roots of the shrubby kinds, or tubers of the ordinary herbaceous Chinese Pæony. The triangular side graft is best, if the tubers employed are of large size, and the splice graft on the roots of the shrubby kinds. The grafting should be done in early autumn, and the worked roots stored in some place where they will not become frozen. New varieties are raised from seed, which should be sown in good soil as soon as ripe. They will germinate the following season, but the entire growth will be directed or expended in producing roots and small tubers; the cotyledons remaining enclosed in the shell of the seed until the spring of the second season. I have raised many hundreds of seedling Pæonies, but never had one appear above ground until the spring of the second season; I have always found them producing roots and tubers the first summer.

Paliurus (Christ's Thorn).—Shrubs of the *Rhamnaceæ* or Buckthorn Family. One species native of Judæa, another of Nepal. The former has rather slender thorny branches, with fruit resembling a head with a broad-brimmed hat on. Not quite hardy in our Northern States, but often cultivated in greenhouses for the sake of its reputed association. Propagated by layers and cuttings of the roots. The latter grow freely if stored in moist sand or moss during the winter.

Passiflora (Passion-Flower).—An extensive genus of herbs and climbing shrubs, bearing large and beautiful flowers of various colors. A few of the larger-growing shrubby species, like the common Granadilla (*P. edulis*), and the large Granadilla (*P. quadrangularis*), produce edible fruit, as large as a lemon, and of a similar form. All the species and varieties easily propagated by cuttings of the young shoots planted in sand, and where they can be given moderate heat.

Paulownia.—A well-known ornamental tree (*P. imperialis*) from Japan. Its leaves are of immense size, rather downy, and heart-shaped. Flowers large, violet-colored in terminal panicles in spring. The flower buds are often winter-killed in the latitude of New York City, but the trees are hardy. Propagated by seed sown in spring, or cuttings of the roots made in the fall and stored in a moderately warm place during the winter.

Phellodendron (Cork Tree).—A small genus of deciduous trees of the *Rutaceæ* or Rue Family, and closely allied to our common Prickly Ash (*Zanthoxylum Americanum*). Two species have been introduced— *P. Japonicum*, supposed to be a native of Japan, and *P. Amurense*, from the Amoor regions in Asia. Propagated by layers and cuttings of the roots.

Philadelphus (Syringa).—A very popular genus of ornamental shrubs, natives of North America, Europe and Asia, including Japan. There are probably less than a dozen species, but a large number of cultivated varieties. Flowers white, and in some of the species sweet scented; others inodorous. The common European varieties have long been known in our gardens under the name of Mock Orange, as the flowers of *P. coronarius* smell somewhat like Orange flowers. All readily propagated by cuttings, layers and suckers.

Phillyrea.—Ornamental evergreen shrubs of the Olive Family, natives of Southern Europe and the East Indies. Flowers small, white and inconspicuous. These shrubs are esteemed for their deep, rich green foliage. Propagated by cuttings and layers.

Photinia.—Evergreen shrubs of sub-tropical countries belonging to the Rose Family. One species is extensively cultivated in warm climates under the name of Loquat tree (*Eriobotrya Japonica* of nurserymen's catalogues), as it bears a yellow edible fruit, resembling small apples. This, and several other species are old inhabitants of our greenhouses. Propagated by cuttings of the ripened wood, and by budding and grafting on the Quince and Hawthorn.

Pinckneya (Georgia Bark).—A small evergreen tree of the Southern States, closely allied to the *Cinchona*, and the bark is supposed to contain the same or similar principles; hence, the common name of the native species, *P. pubens*. Propagated by seeds and cuttings under glass, but must be well supplied with moisture.

Pinus (Pine Tree).—An extensive genus of evergreen coniferous trees and shrubs. Many of the species, and some of the most valuable, thrive in soils unfit for agricultural purposes, it being either too light and dry and thin, or too cold and wet. Pine barrens and Piney swamps are, as a rule, the home of this genus throughout the world, although, upon the whole, it is one of the most useful to mankind. Propagated by seed sown in spring in a half-shady position. The seed should be covered lightly with sandy soil and kept moist until the plants appear, when these must be shaded from the direct rays of the sun, and given water only sufficient to prevent wilting and drooping. The shading should be continued during the entire first season, either with lath screens, as shown elsewhere, or with branches thrown over the seed bed. Varieties are propagated by veneer grafting under glass in August. (See Chapter XVII.)

Pirus (Apple, Pear, Etc.)—A very important genus of the *Rosaceæ* or Rose Family. The common Pear (*P. communis*) and Apple (*P. Malus*), are the most valuable fruits raised in cool climates. The common Quince (*Cydonia vulgaris*) is so closely allied to the Pear that it is used as a stock for many varieties. The same may be said of the Hawthorn (*Cratægus*): although belonging, botanically, to a different genus, they are sometimes employed as stocks for both the Pear and the Apple; but, as I have said in Chapter XVIII., on Selecting Stocks, it is always best to employ stocks as near related to the kind being propagated as possible. In propagating the Pear, the seed should be washed from the ripe fruit, or pomace, while fresh, and before decay has proceeded so far as to have softened the thin shell enclosing the cotyledons or seed proper. It may then be spread out in the shade and dried sufficiently to prevent it from becoming moldy or heating when stored in boxes or bags. Pear seed is usually

SELECT LISTS OF PLANTS.   287

imported dry, but when received it may be mixed with moist sand, and either buried in the open ground or stored in a cool cellar, until it is wanted for sowing in early spring. But it may be mixed with sand as soon as washed from the pomace, and then, buried in the ground, it will germinate more readily and produce stronger and more vigorous plants than seed that has been long dried. The seed should be sown in single or broad drills and covered not more than an inch deep. The seedlings should be taken up when one year old, the tap-root shortened, and the stem cut back to within six or eight inches of the ground, before planting out again. The following August, if they make a good growth, they will be in good condition for budding in the usual way. The next spring they are to be cut back to within four inches of the bud, and all sprouts kept removed from the stock during the summer. The stocks on which the buds have not taken may be splice grafted near the ground, in order to have the rows full of worked trees. The next spring the stump above the bud may be removed with a clean, upward sloping cut. Sometimes one-year-old, and even older, Pear stocks are splice grafted in the winter and planted out in spring, but this mode of propagation is more generally practised with the apple. Apple seed may be treated in the same general way as those of the Pear, but they grow more freely and are far less liable to be injured by rust in summer and other parasitic diseases. Root grafting the Apple, using one and two-year-old seedling stocks, is extensively practised by nurserymen. The stocks should be dug up in the fall and stored in a cool cellar or pit, where they can be readily taken out when wanted for use. Cions should also be cut from the trees of the varieties to be propagated, late in fall or early winter, and stored where they will not shrivel or become softened by water. Well-ripened wood of the present season's growth is used for cions, and that from bearing trees is preferable to shoots from small and immature trees. When ready to commence grafting, the stocks and cions are brought into a warm cellar or room and brushed or washed clean; for if covered with sand or earth of any kind, the knives used will soon become dull, and good work

Fig. 99.
SPLICE GRAFTING THE APPLE.

is impossible with dull tools. Strong manilla paper, coated on one side with wax—or very thin cloth may be used if preferred—is cut into narrow strips for tying in the cions. When the materials are all at hand, proceed with the work by splicing a cion on the crown of the seedling stock, as shown in figure 99, and wind the lower part of the cion and stock with a piece of the waxed paper or cloth, applying only enough to cover the splice. (See Splice Grafting in a preceding chapter). Sometimes nurserymen cut up the long slender roots of seedlings into several pieces, inserting a cion on each ; but better trees and a more vigorous growth will be obtained if the stocks are entire, or only the taproot and some of the lateral ones are shortened, if of too great a length for convenience in handling, in grafting and when planting out. Two or three persons may work together at this root grafting to considerable advantage ; for while one is cleaning and preparing the stocks, another can cut and insert the cion, while the third may apply the waxed cloth or paper, without loss of time in laying down and picking up knives and other implements used in the operation. When the stocks are grafted, they are packed away in moss or soil and stored in a moderately cool cellar, where the process uniting cion and stock will proceed slowly until the time arrives for planting out in nursery rows in spring. The Apple may be readily propagated by budding or grafting in the open air, but root grafting in winter is preferred, because little else can be done in the nursery at this season, and long experience has shown that this mode of propagation answers every purpose, and, upon the whole, is the cheapest. In propagation of the different species of the Mountain Ash (*Pirus Americana*, *P. aucuparia*, etc.,) the seeds should be sown where the young plants can be shaded until they become well established, for in our hot climate the young seedlings are very likely to be burned off, if not protected from the direct rays of the sun. The different varieties may be propagated by budding or grafting on seedling stocks and in the open air.

Pittosporum (Pitch Tree).—Ornamental evergreen trees or shrubs, mostly natives of tropical countries. The most common species in cultivation is the *P. Tobira*, from Japan, bearing small, white, fragrant flowers. Propagated by cuttings of the ripened shoots, planted in sand under glass.

Planera (Planer Tree).—A genus of small deciduous trees closely allied to the Elms (*Ulmus*), of no great beauty, but interesting to the botanist and arboriculturist, as the few species in the genus are natives of widely separated countries ; for instance, United States, Japan and Siberia. Propagated by seed, layers, and grafting on the Elm. (See Chapter XVII.)

Platanus (Plane Tree, Sycamore, Buttonwood).—A very limited genus of only three species, two in the United States (*P. occidentalis* and *P. racemosa*), and one in Europe (*P. orientalis*). Of the latter there are several varieties in cultivation. All large, noble, deciduous trees, but rarely planted in this country, although the European species and varieties are occasionally seen in some of our city parks. Propagated by cuttings of the ripe wood, taken off in the fall, and buried in the ground, and planted in low, moist soil the following spring.

Podocarpus.—A genus of evergreen trees of the *Taxaceæ* or Yew Family, mostly native of warm climates ; one species—the *P. japonica*—nearly hardy in this latitude, but it is doubtful, if it will prove of much value as an ornamental tree at the North, except in sheltered situations.

Propagated by cuttings of the mature shoots set in cold frames, or in a cool greenhouse during the winter months.

Poinsettia.—See *Euphorbia.*

Populus (Poplar, Aspen, Cottonwood).—A genus of about twenty species, one-half the number indigenous to North America, and one species—the Quaking Aspen (*P. tremuloides*)—extending northward to the Arctic Ocean. With few exceptions the Poplars are large, rapid-growing, deciduous trees, thriving in a great variety of soils, but succeed best in one that is rather moist. Extensively cultivated in cold climates for ornament, fuel and shelter. The wood is rather light, of little value for uses, where it is exposed to the weather, but is valuable for fuel, especially where better kinds cannot be obtained. Propagated by seeds, suckers, cuttings of the branches and roots. Varieties are usually propagated by cuttings, or by budding and grafting upon stocks of the free-growing species. The seeds are small and produced in pendulous catkins, appearing before or at the time of the unfolding of the leaves. The seeds ripen early in the season, or about two months from the time the flowers appear, and if sown as soon as ripe, they will germinate and produce plants a foot or more in height the first season. Seeds should be scattered over the surface of the seed bed and merely raked in, or a little fine soil sifted over them. If no rain falls soon after the seed is sown, water must be given in liberal quantities until the plants appear and become well established. Cuttings may be made of either one or two-year-old wood, and planted in the fall or spring.

Potentilla (Five Finger, Cinquefoil).—An immense genus of the Rose Family, and with few exceptions herbaceous perennials, the species being widely distributed around the world in the Northern Hemisphere. The Shrubby Cinquefoil (*P. fruticosa*), is a low-growing shrub, two to four feet high, with bright yellow flowers; the plants usually continuing in bloom from early spring until checked by frosts in autumn. This handsome little shrub is indigenous to the colder regions of both North America and Europe, but thrives in much milder climates. Propagated by divisions of the clumps, by layers, and cuttings of the mature wood, taken off in the fall, and stored in the usual way until spring.

Prunus (Plum, Prunes).—In most of the recent botanical works the Plum and Cherry are placed in the genus *Prunus*, but the propagator of these fruits is obliged to keep each group distinct or separate, as the true Cherries (*Cerasus*), and the Plums (*Prunus*), cannot be, except in rare instances, interchanged in propagating by budding and grafting, while the Peach, Almond, Apricot and True Plum are so closely allied that any and all of the species and varieties may be employed indiscriminately as stocks for one and another. Still, there is always a preference when selecting a species or a variety for a stock. (See Fruit Stocks, Chapter XVIII.) Propagated by seed, layers, cuttings of the mature wood, and cuttings of the roots. The latter mode is not recommended, as most of the species and varieties are inclined to produce suckers rather too freely. Some of the varieties of the common European Plum (*P. domestica*), are readily increased by cuttings of the ripe wood treated in the same manner as usual with the Currant and Gooseberry. The Myrabolan Plum, so largely used in France for stocks, may be propagated by cuttings; also many of our choice garden varieties. The seeds or Plum stones should be placed where they will be kept moist and cool during the winter months, and if they freeze while moist, the shell will open all the more readily when planted out in the spring.

Pseudotsuga (Douglass Spruce).—A species of conifers found in the Rocky Mountain regions; closely allied to the True Spruces (*Picea*). It is a hardy tree, thriving in our Northern Atlantic States. Propagated by the same modes as the more common species of the Spruce.

Ptelea (Hop Tree).—A small genus of North American shrubs or small trees. Two species are native of the United States; one, the *P. trifoliata*, is common in the Middle and Western States, and the other one, *P. angustifolia*, from Texas westward to California. The broad, oblong winged seed is sometimes used as a substitute for the common Hop. Propagated by layers and seeds sown in autumn, or preserved in sand until spring.

Pterocarya (Winged Walnut). A small genus of deciduous trees from Asia, closely allied to the Hickories (*Carya*) and Walnuts (*Juglans*). Propagated by seed and layers, also by suckers that usually spring up about the main stem, these producing roots sufficient to admit of allowing them to be taken off and planted out with safety.

Pterostyrax (Winged Storax).—An ornamental deciduous shrub or small tree from Japan, bearing creamy-white, fragrant flowers. It belongs to the Storax Family (*Styracaceæ*), and is closely related to the Halesias. Propagated by seeds, layers, and by grafting on the Halesia. Veneer grafting under glass in late summer is the most certain mode.

Punica (Pomegranate).—This is one of the few fruits that appears to have come down to us from very ancient times, and almost in its primitive condition. The Promegranate tree is of a rather bushy habit, growing from twenty to thirty feet high in tropical countries, although it is readily controlled by pruning, and may be trained in the form of a small shrub. It is extensively cultivated throughout the semi-tropical and tropical countries of the Old and New World, and highly prized as an ornamental and fruit-bearing tree, in our Southern States. The fruit is as large as an ordinary apple, and the numerous seeds imbedded in or surrounded with a juicy pulp; it is used in hot climates for making cooling drinks. There are several varieties in cultivation, all readily propagated by seeds, layers and cuttings. Very scarce varieties are sometimes propagated by grafting on the more common sorts.

Quercus (Oak).—A very large genus of evergreen and deciduous trees and shrubs. About forty species are found within the limits of the United States, and more than two hundred additional species are indigenous to other countries of the northern hemisphere. They are for the most part highly valued for ornament as well as their hard and durable timber, and there are few other kinds of trees that have been more celebrated in peace and war than the Oaks. The flowers are monœcious: the staminate flowers in catkins, the pistillate in a cup-like involucre, covered with scales; the seed a one-celled nut, well known under the common name of Acorn. The acorns of some species of the Oak ripen the first season; in others not until the autumn of the second, but all appear to be inclined to germinate very soon after they have fallen and come in contact with the moist earth; consequently, whatever is to be done in the way of gathering, storing, or sowing, must not be delayed long after the acorns begin to fall from the trees. If the acorns are shaken from the trees, or picked up as they fall, they may be preserved in a cool room for weeks, and some species for several months, without serious injury; but, as a rule, acorns are rather difficult to preserve in good condition for growth, and the sooner they are sown after ripening, the better. In some few of the species, the nuts do not fall out of the cup, but both drop together, and the acorn remains within the husk until it germinates the following spring, when it bursts both the inner and outer shell in its germination. But this form of acorn is rather an exception than the rule; those of a larger majority of the species begin to grow in the fall, the root or radicle penetrating the soil for several inches, thereby holding the acorn in a position for the production of the plumule or stem, the following spring. Any one who has taken a stroll on the edge of an oak forest late in the fall, must have noticed these "anchored" acorns, while the cotyledons or seed-leaves still remained within the inner shell. But when fairly within an oak forest, we find that the acorns as they fall do not come in direct contact with the soil, the layers of old tough leaves on the surface preventing; consequently the larger proportion of the acorns perish for the want of suitable anchorage, or conditions favorable for growth.

In sowing acorns, they may be scattered in single or broad drills, or even broadcast over the surface of a seed-bed, and then lightly covered with hay, chaff, or very fine old manure or leaf-mold. In such positions they will take root in the autumn, and the next season make a vigorous growth. As with other nuts, a light, sandy soil will insure a far greater number of fibrous roots than a heavy one. Those persons who live near oak forests can always secure a stock of seedlings, without the trouble or cost of gathering and sowing the nuts, by merely raking away the old leaves from under the trees of the species they desire to secure. The acorns falling on the bare ground will soon sprout and become fixed in position, and the leaves falling later will give them ample protection. I have practised this with eminent success with our common White and Black Oaks, and while the seedlings obtained in this way were not as large as those raised in the nursery, they were still fair plants and made a good, vigorous growth when transplanted to nursery-rows. Of course, it is not to be supposed that seedlings can be obtained in this way in regions where hogs are pastured in the woods. Rare species and varieties of the Oak are propagated by grafting. (See Chapter XVII.) But it will be found in practice that the Oaks are rather difficult subjects to deal with, and grafting in the open air should be performed early in spring. Veneer grafting in August, under glass, is the most likely to be successful, especially with the evergreen and variegated-leaved Oaks.

Raphiolepis (Indian Hawthorn).—A genus of low-growing evergreen shrubs from China and Japan. They are closely related to the Hawthorns (*Cratægus*), with large, thick, dark-green, leathery leaves and white or pink, sweet-scented flowers in pyramidal-shaped clusters. The Japan species (*R. Japonica*) is said to be hardy in the gardens of London, England, and will probably succeed here with a slight protection in winter. Propagated by cuttings of the half-ripened shoots planted in a greenhouse or in close frame in the open ground.

Retinispora.—See *Chamæcyparis*.

Rhamnus (Buckthorn).—A very large and widely distributed genus of evergreen and deciduous trees and shrubs. There are a half dozen indigenous species, and the common European Buckthorn (*R. catharticus*), formerly used as a hedge-plant in this country, has run wild in many places. The berries of this species were formerly used in medicine, and the juice of the ripe berries yield a pigment called "Sap-green," in common use by water-color painters, while the bark of the branches and roots yield a valuable yellow dye. Propagated by seed, and these, being quite hard, like the Hawthorns, do not usually germinate until the second season, and require the same treatment.

Rhododendron (Rose-Bay). — An extensive genus of evergreen shrubs and small trees, with large, and usually very thick, smooth, green leaves. Taking them as a whole, the Rhododendrons may be placed in the front rank among the most showy and elegant of ornamental plants. There are not only a large number of species, but almost innumerable varieties in cultivation, while new ones appear every year, as hybridizing and crossing is readily effected, resulting in wide variation in habit of plant, and in form, size, and color of the flowers. Propagated by seeds, layers, and grafting, and some of the more slender and tender species by cuttings of the young shoots, removed with a heel of the old wood, then plunged in sand in close frames with bottom heat. The seeds are very minute, and must not be covered deeply. They are usually sown in shallow boxes or seed-pans, filled with a mixture of leaf-mold and clean sand, and then placed in a close frame until the plants appear, care being required in watering, lest the seeds are washed out on the surface. When the plants are large enough to handle, remove into other boxes or pans, giving them a little more room for growth, then replace in the frames for a few days, or until the plants become established in their new position; after this they may be gradually hardened off by removing the covers of the frames for a few hours at a time. Hardy species and varieties may be planted out in a sheltered position, when a year old, although there will be nothing lost by keeping them in the seed pans a little longer. There are several modes of grafting practised with Rhododendrons, but veneer grafting or side grafting in August and September under glass are the most certain as well as the most convenient. (See Chapter XVII.)

Rhus (Sumac, Smoke Tree, Etc.)—A very large and important genus of evergreen and deciduous trees and shrubs. Some are extremely poisonous, others yield astringent properties valuable for tanning Turkey and Morocco leather. One Japanese species is said to yield the famous lacquer, another a valuable wax, and taking it all together, the genus is a valuable one in the arts, besides those species cultivated for ornamental purposes, like the common Venetian Sumac (*R. cotinus*), and the rather rare but closely allied American Smoke tree (*R. cotinoides*). Propagated by seeds, layers, cuttings of the roots, and some species by cuttings of the ripened wood, made in autumn before the branches have been severely frosted.

Ribes (Currant, Gooseberry).—Well-known, berry-bearing, deciduous shrubs, mostly native of cold climates, succeeding very poorly or not at all in warm ones. New varieties are raised from seeds, which germinate at a very low temperature, and for this reason they must be stored in a very cool place, else they will sprout in spring before the weather will permit of a continuation of growth. Seeds should be washed from the pulp, then mixed with pure sand and placed in the shade out-of-doors; on the north side of a building is usually a safe position to prevent early germination in spring. If only a few plants are to be raised from seeds, they can, of course, be started under glass, but the open ground is preferable, giving the young plants shade until they are well established. All the species and varieties are readily increased by layers and cuttings of the mature one-year-old shoots. The usual practice is to make the cuttings of Currants early in the fall, and plant immediately, protecting the cuttings with a mulch of coarse hay or manure. They will produce roots before the ground freezes, and the mulch will prevent lifting by frost. Gooseberry cuttings may be made a little later in the fall, but not planted out until spring. (See Grafting, Chapter XVII.)

Robinia (Locust Tree).—A genus of few species, the most important of which is the common Locust or False Acacia (*R. Pseudacacia*), a large, deciduous forest tree, with deep green pinnate leaves and loose-drooping racemes of white fragrant flowers. The timber of this tree, when of slow growth, is one of the most durable known. Formerly, this tree was extensively planted in the Eastern States, and is still to a limited extent; but of late years the Locust borer (*Cyllene pictus*), has been so destructive to the trees that very few are now planted. There are two other indigenous species—one a mere shrub with rose-colored flowers—that are cultivated for ornament. Propagated by seeds, sown in the fall or spring, and the ornamental species and varieties by seeds, cuttings of the roots, and by budding and grafting. (See Stocks, Chapter XVIII.)

Rosa (Rose).—A very extensive or limited genus, according to the botanical authority one consults for information on this point. Some make the number of species 250, but modern botanists have reduced it to about thirty, all natives of the temperate and colder regions of the Northern Hemisphere. The number of varieties in cultivation is unknown, for old ones become obsolete, as new ones are introduced, although there are always several thousand enumerated in rose-growers' catalogues. The Rose is a universal favorite among all civilized nations, and the difference in the fragrance of the flowers of the varieties is only equalled by their variation in size, form and color. To attempt to classify the cultivated Roses at the present day would be a hopeless task, for species have become so intermixed that specific characteristics have been mostly obliterated. In the propagation of the Rose, every mode at all applicable to ligneous plants is employed in its multiplication. Roses are raised from seeds, not only for the purpose of producing new varieties, but sometimes for stocks on which to bud or graft the improved sorts. For all the common, hardy varieties of the Rose that produce seed, the fruit or "heps"—as English gardeners call them—should be gathered when ripe, and thrown into some vessel where they will be moist until the surrounding pulp becomes soft ; then crush and wash out the seeds, and either mix with sand and set aside where they will freeze, or sow immediately in a bed in the open air, and treat as recommended for Hawthorn and other similar seeds. But if there is danger of mice getting into the bed during the winter, it is best to keep the seeds in the boxes with sand, covering with wire netting to keep them out. In spring, sow the seeds and sand together in seed-pans, boxes, or in an outside frame, but always where they can be given plenty of water and be protected from vermin. Scalding the seeds before sowing will hasten germination, but it is not usually necessary, if the seeds have been kept moist and cold during the winter. Sometimes the seeds will not sprout until the second year, and it is well to keep the seed beds moist throughout the summer, even if some plants do appear the first season, as more will usually come up the second. As soon as the plants are large enough, they should be carefully lifted and transplanted into other frames or boxes. The tender Roses may be raised in the same way, only avoid subjecting the seed to as low a temperature, and it is better to sow it in pans or shallow boxes in the house. Green cuttings of what are called the Tea, Noisette, Bourbon, Hybrid and Hybrid Perpetuals, and several other classes, strike root quite freely in sand under glass, and this is the usual method of propagating these varieties. Some of them, however, are rather slow growers and shy bloomers on their own roots, and to increase the growth and vigor, they are budded on hardy and strong-growing varieties, such as the Manetti, Sweet-briar and Dog Rose. These kinds are also employed as stocks for many other varieties, both tender and hardy. Hardy Roses, however, if naturally of a free-growing habit, are to be preferred on their own roots, especially for amateurs, who cannot be always on guard, lest some sucker from the root comes up and robs the graft of its nutriment. Roses are usually budded in the open ground in summer, and at any time when good plump buds can be obtained, and the stocks are in condition to receive them. Grafting is rarely practised in the open air, but splice grafting on pieces of roots of some common variety is a convenient and rapid mode of increasing rare varieties, the cuttings of which do not strike root readily. Cions of one bud will answer, but if the wood is short jointed, two are better; the grafted roots being planted in boxes filled with sand or very light soil, and afterwards given the same care as ordinary cuttings. This is a common mode of propagating Moss Roses and other similar hard-wood varieties. Most of the Climbing Roses, of both native and foreign origin —also the classes known under the name of Perpetual and Hybrid Perpetuals—are quite readily propagated by ripe wood cuttings, placed in a cool greenhouse in autumn, or in protected frames in the open ground. They are also readily increased by green cuttings taken from plants in the open air, or from those forced under glass; the short, spur-like shoots, taken off with a heel, are preferable for this purpose to the more vigorous and succulent wood. But the most simple method of propagating Roses is by root cuttings, and there are very few species or varieties that cannot be readily multiplied by cuttings of their roots. Some species and varieties—like the Moss Roses, Briars and common June Roses—usually considered difficult to propagate by layers and cuttings of the shoots, grow very freely from pieces of their roots, if these are given sufficient time to develop adventitious buds before attempting to force them to produce new roots and stems. The time to make root cuttings of hardy Roses is in the fall, as soon as the plants have been checked by cool weather. The roots, or a part of them, may be removed without digging up the entire plant, but it is better to lift the plants, following out the roots to their ends, then cut away all those suitable for cuttings—the larger roots make the best, but those of not more than one-sixteenth of an inch in diameter will answer—and cut all up into pieces of two to three inches in length. Pack these pieces between layers of damp moss—the common Sphagnum from the swamps and low grounds is the best, but if this cannot be obtained, pure clean and sharp sand may be used instead. These root cuttings may be packed in well-drained boxes or large flower pots, or any similar vessel, but in all cases they should be well drained and absolutely clean and free from any taint or substance likely to generate or promote the growth of mildew and mold. These boxes or other vessels containing the cuttings, may be buried in a dry place in the open ground or set away in a cool cellar, where they can be examined from time to time during the winter, for the purpose of ascertaining their condition and giving water, if it should be required. If buds push too rapidly, lower the temperature; and if they do not come forward as rapidly as is thought necessary, increase the temperature, or remove to a warmer place. Root cuttings of some varieties will push into growth under exactly the same conditions, where others will remain quite dormant, and for this reason it is well to place the cuttings where they can be examined. All that is necessary or desired is to secure the development of one or more buds on each cutting by the time the weather will permit of planting them in the open ground in spring. These root cuttings may be sown in drills and covered with good rich soil to the depth of two inches, and water applied to settle the ground, or it may be packed slightly with the back of a hoe or light roller. Good strong plants are usually produced from such root cuttings the first season. With the more delicate, tender varieties, like the Teas and Bourbons, the roots should be treated as recommended for those of the Bouvardia—which see.

Rubus (Raspberry, Blackberry).—A large and interesting genus of plants, the species pretty widely distributed over the world. Some of the species are strong, large, upright shrubs with perennial woody stems; in others—as with most of our cultivated species and varieties—the stems are biennial—that is, growing one season, fruiting the next, and dying down in the latter part of summer or early autumn. The few herbaceous species are natives of cold climates, while the evergreen are mostly indigenous to warm or tropical ones. In the propagation of these fruits, seed is seldom employed, except for the purpose of producing new varieties, and it may be sown as soon as taken from the ripe berries, or the latter may be dried, and the seeds preserved in good condition for several years, and when wanted for sowing it is only necessary to soak the dried fruit for a few hours in warm water, wash out the seeds, and sow in good soil, watering the bed freely until the plants appear and are large enough for transplanting. The ornamental Brambles, as represented in the Atlantic States by the Purple-Flowered Raspberry (*R. odoratus*), and in the Western and Pacific Coast regions by the Salmon-Berry (*R. Nutkanus*), and one or two other closely allied species, and from China by the Rose-Flowered Raspberry (*R. rosæfolius*), are rarely cultivated for their fruit, although it is edible, but rather deficient in flavor. The species most valued for their fruit belong to the two groups known as true garden Raspberries and Blackberries. There are upright-growing and trailing species, and varieties in both groups. Those with a trailing habit increase naturally by a natural process, called rooting, or taking root at the tips, the long, flexible canes bending over and taking root, as seen in the varieties of the Black-caps (*R. occidentalis*), among the Raspberries, and in the Low Blackberry or Dewberry (*R. Canadensis*). The trailing species do not produce suckers, but sometimes numerous sprouts spring up from around the base of the main stems, and the old stools may be taken up and divided into several plants, when such a mode of propagation is desirable for the more rapid increase of a variety. The upright-growing species of both groups produce suckers more or less freely; these latter are employed in their propagation, and taken up in spring or fall, and set out to make new plantations. But in the propagation of the garden Blackberries, better plants may be raised by root cuttings than are produced naturally in the form of suckers, and the same is true of such varieties of the Raspberry, as the Purple Cane, Philadelphia, and Shaffer's Colossal, because when raised from root cuttings the plants are far better supplied with small fibrous roots, which are readily preserved in the digging and transplanting. The method of propagating the different species of *Rubus* by root cuttings is the same as described for Roses and other similar hardy plants, but the roots of some of the varieties grow much more freely than others. The ornamental varieties, especially those cultivated under glass, are readily propagated by green cuttings taken off close to the old wood, or with a heel, and planted in sand in a close frame.

Salisburia (Ginkgo, Maidenhair Tree).—A large, hardy, deciduous ornamental tree from Japan and China. It belongs to the *Taxaceæ* or Yew Family, and is the only representative of the genus. The pistillate and staminate flowers are on different trees, the pistillate flowers solitary; fruit drupe-like, with a large nut-like seed. Although introduced more than a century ago (1784), the Ginkgo is still far from being a common tree in this country. Propagated by seeds imported mostly from oriental countries, by layers, and cuttings of the young shoots taken off with a heel in midsummer, or of the ripened twigs in autumn, and planted in frames in a greenhouse. Varieties are propagated by the same modes, or by grafting in the open air early in spring, or under glass in August.

Salix (Willow, Osier).—An immense genus of widely distributed trees and shrubs, all thriving best in moist soils and swamps. The larger number are so readily propagated by cuttings that other modes are seldom practised, except to produce small, weeping trees, by budding or grafting the small, low-growing species on stocks of those of an upright habit. (See Stocks, Chapter XVIII.)

Sassafras.—A well-known native deciduous tree of the Laurel Family, with very fragrant foliage, and roots with thick, yellow, spicy bark. A handsome tree, but produces suckers far too freely for admission into cultivated grounds. Readily propagated by seed and cuttings of the roots. The species is *S. officinale*, and, in some works, *Laurus Sassafras*.

Sambucus (Elder).—A small genus of deciduous shrubs and a few herbaceous plants. The shrubby species propagated by seeds, cuttings, and layers, and the herbaceous by division of the roots. The shrubby species usually increase far too rapidly by suckers, and often become a nuisance in grounds of limited extent.

Sciadopitys (Umbrella Pine).—A rare coniferous evergreen tree from the mountains of Japan, where it grows from one hundred to one hundred and fifty feet high. A very distinct and hardy conifer, but apparently of rather slow growth. Usually propagated by seeds imported from Japan, but the seedlings make an exceedingly slow growth, at least during the first half dozen years. Cuttings of the half-ripened shoots, taken off in summer and planted under glass, strike root quite readily; and where one has good stock plants to supply the cuttings this is the most expeditious mode of propagation.

Sequoia (Great Tree of California, Etc.)—A genus of two species of coniferous evergreen trees both native of California. One species, the

Sequoia gigantea, is supposed to grow to a larger size than any other tree indigenous to Europe or America, and is only excelled in size by a few species of the *Eucalyptus* in Australia. Propagated by seeds, which should be sown in a half-shady position, lightly covered, and kept moist until the plants appear; then some care is required to prevent damping off in warm weather, although watering must not be neglected. Only moderate heat is necessary to insure a healthy growth. A temperature of from fifty to sixty degrees will prove to be far better than a higher one. These trees may also be propagated by layers and cuttings, treated the same as usual with the common Arbor-vitæ, Yews and Junipers. The *Sequoias* thrive best in a rather moist soil, and in what is usually termed a moist, cool climate. *S. sempervirens* is the valuable Red-wood.

Sheperdia (Buffalo Berry).—A genus of three North American deciduous shrubs or small trees, found only in the cool regions of the Northwest. The largest-growing and most valuable species (*S. argentea*), is found in northern New Mexico and northward to Alaska; in the latter country, I am informed by correspondents, the fruit is gathered in immense quantities by the Indians for use in winter. The fruit is of a bright scarlet color, resembling small currants, juicy, rather acid, but pleasant flavored. This species succeeds perfectly in this latitude; I raised plants from seed gathered on the Upper Missouri thirty years ago; they are still alive; and the pistillate plants seldom fail to bear a good crop of fruit. The flowers being diœcious, it is necessary to have trees of both sexes growing near together, in order to insure the production of fruit on the pistillate or female plants. Propagated by seeds, which should be removed from the ripe fruit, and either sown in the fall, or mixed with sand, kept cool and moist during the winter, and then sown in the spring in a half-shady place, as the young seedlings are rather sensitive to the direct rays of the sun in this climate.

Skimmia.—A genus of evergreen shrubs from Japan, closely allied to the Evergreen Barberries (*Berberis*), but with white, sweet-scented flowers, succeeded by bright red berries. Not quite hardy in this latitude. Propagated by layers, or cuttings planted under glass.

Sophora.—A genus of about twenty-five species of leguminous trees, shrubs or herbs, mostly native of warm countries. There are two or three evergreen shrubby species found in Texas and Mexico, and one in Arizona and California. The species most highly valued in cultivation is the *S. Japonica* and its numerous varieties, all moderately hardy in our Northern States. The latter are propagated by seeds, layers, and by grafting. The pendulous-branched and variegated-leaved varieties are grafted on stocks of the species in the open air in spring. Tender varieties and species may be increased by cuttings taken from plants forced under glass.

Spiræa (Meadow-Sweet, Bridal Wreath).—A genus of a half hundred species, mostly native of temperate regions of the Northern Hemisphere. They are mostly low-growing deciduous shrubs, a few with persistent evergreen leaves, and others perennial herbaceous plants. About a dozen species are natives of the United States, two of which are low herbaceous plants of the Rocky Mountain regions and far northward. Another herbaceous species of tall growth—viz., *S. Aruncus*—has dioecious flowers, the male plant being the one most common in cultivation. This species is very widely distributed and found growing wild in the Alleghanies, thence northward to Alaska and through Northern Asia and Europe. The common shrub known by the name of Nine-Bark, and in most botanical works as *Spiræa opulifolia*, is now placed in the genus *Neillia* Don., along with four or five other species found in the mountains of Asia. The shrubby species are propagated by cuttings of the ripe wood, by layers, and cuttings of the roots. Herbaceous species by division of the clumps.

Staphylea (Bladder-Nut).—A small genus of three or four species of large shrubs with small white flowers; the seeds produced in a three-lobed, three-celled bladdery pod. Our native species (*S. trifoliata*) is common in low grounds in the Northern and Western States. The European Bladder-nut (*S. pinnata*), and the Japanese species (*S. bumaldi*), are occasionally cultivated. Propagated by seeds, layers, suckers and cuttings of the large, rather fleshy roots.

Stuartia.—Large, hardy, deciduous shrubs, with showy white flowers, resembling those of the Tea-plant. There are two species indigenous to the Southern States, and one or two to Japan; the latter have recently been introduced. Propagated by seeds and layers, but the latter do not strike root very readily. It is said that they are propagated by ripe wood cuttings, planted under glass, in European nurseries, but I have no experience with this mode.

Styrax (Storax).—A genus of very ornamental deciduous shrubs, represented by a half dozen species in the United States. The European species (*S. officinale*) is noted for producing the very powerful and fragrant balsam, known by the name of Storax. Propagated by seeds and layers, and by grafting. (See Halesia, and Selection of Stocks, Chapter XVIII.)

Symphoricarpus (Snowberry, Indian Currant).—A genus of about a half dozen species of low-growing North American shrubs, cultivated for their ornamental berries. The hardy species are common in gardens. Propagated by cuttings or suckers; the latter produced in great abundance.

Symplocos (Sweet-Leaf, Horse-Sugar).—Evergreen trees or shrubs of the Storax Family. One species native of the Southern States (*S. tinctoria*), one of Mexico, and two or three in China and Japan. Propagated by cuttings under glass.

Syringa (Lilac).—A genus of old and well-known ornamental shrubs. There are but few species, but an immense number of cultivated varieties, and new ones are being brought forward almost every year. Propagated by seeds, suckers, layers, and cuttings of the larger roots. Also by budding and grafting in the open air in summer or early spring. The Lilacs will grow when grafted on the Ash and common Privet, but the union between the two kinds of wood is seldom perfect, and the plants not very long lived.

Tamarindus (Tamarind).—Semi-tropical or tropical evergreen trees bearing delicious fruit, well known in all great cities in its preserved or dried state. Readily propagated by seeds, which should be sown in a hot-bed or where it can be given bottom heat, and the young plants taken up and potted singly when a few inches high, or in hot climates set out in a half-shady bed. The plants may also be increased by cuttings planted in sand under glass, or in the open ground in tropical countries.

Tamarix (Tamarisk).—Slender-growing shrubs or small trees from Europe and Asia. All are evergreen in warm climates, or when grown under glass; but the French Tamarisk (*T. Gallica*), the species most common in the gardens of this country, is slowly deciduous in the latitude of New York City and a few degrees further south. The hardy species and varieties are readily propagated by cuttings of the mature shoots, planted in the open ground, either in the fall or spring.

Taxodium (Bald Cypress).—A small genus of lofty-growing, deciduous coniferous trees, with very short and narrow, or long and slender, thread-like leaves. The Bald or Deciduous Cypress is a familiar indigenous representative of the genus growing abundantly in the swamps of the Southern States. Although this tree is a native of a warm climate, still it is quite hardy in most of our Northern States, and thrives in almost any kind of soil. The *Glyptostrobus* of China are now considered as only species of the genus *Taxodium*, although the former name is still retained in nurserymen's catalogues, and in a few botanical works. The *Taxodiums* are readily propagated by seeds, treated in the same manner as those of the ordinary conifers. Also by layers, and the cuttings of the young shoots in summer, placed in pure sand, constantly saturated with water. They will also strike root in water alone, but the sand is preferable, because it will hold each cutting in one position or place, until the roots are formed. The weeping, variegated, and other varieties of the oriental Taxodiums, may be grafted on stocks of the American Bald Cypress. Grafting in spring in the open air and close to the ground is sometimes practiced with success, but the cions should be shaded for a time with paper caps, or an inverted flower pot will answer. Veneer grafting under glass in August is however the preferable mode for all kinds of coniferous trees.

Taxus (Yews).—Well-known ornamental evergreen trees and shrubs. They are closely allied to the conifers, but the fruit is not a cone, but drupe-like, and the seed enclosed in a soft, bright red, cup-shaped berry. Our American Yew (*T. Canadensis*) is a low, prostrate shrub, found common in woods far north. Propagated by seeds, layers, and cuttings of the green twigs under glass, or the mature wood taken off in the fall and planted in frames, where they can be given slight protection during the winter.

Tecoma (Trumpet Creeper).—For propagation see *Bignonia*.

Thea (Tea Plant).—See *Camellia*.

Thuja (Arbor-Vitæ, White Cedar).—A very extensive genus of coniferous evergreen trees, or a very limited one, depending entirely upon whose classification we adopt as our guide. If the *Retinisporas*, *Chamæcyparis*, *Libocedrus*, and *Biotos* are excluded, as they are by some botanists, a very limited number of true *Thujas* remain—or only two, and these are both indigenous to the United States. But if all the species of the above so-called genera are included, then the genus will be a moderately large one. But the different species are so closely allied that one mode of propagation answers equally well for all. They are readily propagated by seeds, sown in spring in half shade, and watered as often as necessary to keep the surface of the bed moist ; by layers, cuttings of the green shoots under glass, and of the ripe twigs in autumn, planted in frames out-of-doors, or in a cool greenhouse. Grafting under glass is extensively practiced in increasing the plants of rare species and varieties. (For Stocks, see Chapter XVIII.)

Tilia (Linden, Basswood).—A small genus of only about a half dozen species of large, deciduous trees of the temperate regions of Europe and America. All valuable for ornament and for their timber. Propagated by seeds, layers, budding and grafting. The seeds should be sown as soon as ripe, or packed away in moist sand until spring, and then sown. Sometimes the plants will not appear until the second season. Layers strike root quite readily, if notched or merely twisted until the bark is slightly broken. Budding and grafting in the open ground is the usual mode of increasing rare varieties. (See Chapter XVIII., on Selecting Stocks.)

Torreya (Stinking Cedar).—Evergreen trees of the Yew Family. One species, native of Florida (*T. taxifolia*), is called "Stinking Cedar"; another, on the Pacific Coast (*T. Californica*), is known under the common name of "California Nutmeg." There are in addition several oriental species. Propagated by the same modes as the common Yew.

Ulmus (Elm).—A genus of noble, ornamental and useful, deciduous forest trees, mostly natives of Europe and America. There are but few species, but a large number of natural varieties, as the elms are noted for their wide variation from what may be considered the normal types. Long cultivation and the raising of immense numbers from seeds under artificial conditions, has still further augmented the number of distinct varieties. Seeds of most of the species ripen early in summer, and should be sown as soon as they fall from the tree, and lightly covered with soil. Some will germinate in a few days, while others remain dormant until the following spring. Varieties are readily propagated by layers and by grafting on strong stocks of closely allied species in the open air. (See Stocks, Chapter XVIII.)

Vaccinium (Cranberry, Huckleberry, Blueberry, Etc.)—A large genus of hardy, upright-growing shrubs, or low, trailing vines. A few are evergreen, but they are mostly deciduous, and the larger number natives of North America. The low, trailing Cranberries, *V. Oxycoccus* and *V. macrocarpon*—especially the latter and its many varieties—are extensively cultivated, growing freely in low bogs and swamps, and the fruit is well known in our markets. They are readily propagated by cuttings planted in wet soils. But the Huckleberries and Blueberries are rarely cultivated, although there is really no good reason why they should not be, for the plants thrive on almost any light soil, and even those species which naturally grow in swamps and peat bogs will thrive on high and dry soils, provided they are light and sandy. The shrubby species are propagated by seeds and layers. The seeds being very small, they should be sown in shallow boxes filled with vegetable mold and sand, and but lightly covered with fine shreds of moss from the swamps, and this kept constantly saturated with water, until the plants appear, after which only water sufficient to keep the soil moderately wet need be applied.

Viburnum (Arrow-Wood, Cranberry Tree).—A large genus of elegant ornamental shrubs, a few bearing edible but not very highly prized fruit. The common Snowball tree is only a variety of the wild Cranberry tree (*V. Opulus*) of the swamps of our Northern States and Great Britain. The Chinese Snowball (*V. plicatum*) is a more recent introduction, and even more highly prized than the older favorite. Propagated by layers; some of the species by cuttings of the mature wood; others, like the Chinese Snowball, by cuttings of the young, immature shoots taken off with a heel and planted in sand under glass.

Vitex (Chaste Tree).—A genus of tropical and sub-tropical evergreen and deciduous shrubs and trees. They are rarely seen in cultivation, although the Chaste tree (*V. Agnus-Castus*) from Southern Europe, is sometimes found in old gardens in the North, but more common South. The Chinese Cut-leaved Chaste tree is also occasionally cultivated, and is very nearly hardy in the latitude of New York City. Propagated by layers, or ripe wood cuttings planted in a sheltered position in autumn.

Vitis (Grape).—An important genus of climbing shrubs, bearing edible fruit in clusters. The number of species undetermined, as the opinions of botanists differ in regard to the value of the variable specific characteristics for the purposes of classification. The European varieties of the grape are supposed to have descended from one species—viz., *V. vinifera*—and the cultivated American varieties from several indigenous species, through natural variations, or by crossing and hybridizing under domestication. There are also hybrids between the European and American species, and various grades of intermixtures of species and varieties. Propagation by seeds is mostly practiced for the purpose of producing new varieties—which is not at all difficult—from any of the improved cultivated varieties. Seeds from black varieties—like the Con-

SELECT LISTS OF PLANTS. 303

cord—sometimes give pure white or greenish-colored varieties, even in absence of any attempt to cross-fertilize the flowers. Still, it is always best to employ artificial fertilization where hybrids or any intermixture of varieties are desired; for, if such operations are left to chance, the results are very uncertain. Grape seeds should be removed from the fruit when ripe, and then mixed with sand and preserved in a moist condition until spring, then sown in boxes under glass or in the open ground, and covered about one-half inch deep. For the hardy species, the seeds will germinate more freely if placed where they will freeze during winter than if stored where no frost will reach them. Mice are usually so abundant, and so fond of grape seeds, that it is seldom safe to sow them in the fall, otherwise this would be the best plan, and for all kinds. The seedlings should be transplanted into nursery rows when a year old, and the plants set out about four feet apart each way and trained to a single cane and a stake until they bear fruit, then those worth preserving may be increased by any of the usual methods of propagation and the others destroyed. But with the most careful selection of parent plants to raise seedlings from, the chances are not more than one in a thousand of obtaining a variety superior —or even equal—to the best of those already in cultivation. I do not say this to discourage any one who has the inclination and time to spend in making experiments in this direction, for we need far better varieties of the native Grape than any we now possess; but multiplying varieties without a corresponding advance in the intrinsic merits of those produced has already gone far enough—in fact, too far, for the general good of this branch of horticulture. The most common modes of propagating the grape are by cuttings, layers, and grafting. In making what are called ripe wood cuttings, the past season's growth is used; that is, shoots that have been produced during the summer are taken for cuttings in the fall. That which is strong, vigorous and well ripened is to be preferred, but overgrown canes do not always make the best cuttings, or strike root as readily as those of medium size. The best length for cuttings is a moot point. Among vineyardists in Europe they use very long cuttings, while we prefer much shorter ones, and think that we know that

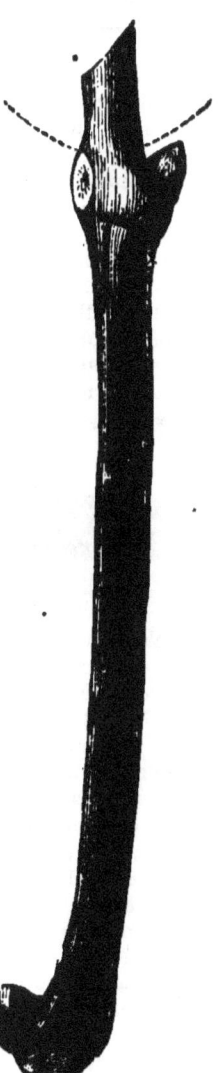

Fig. 100.
GRAPE CUTTING.

they are the best. A cutting of six or eight inches in length, when properly planted, will produce as good or better plant than one twice that length. Furthermore, no modern scientific horticulturists would plant cuttings in a vineyard where the vines are to remain and bear fruit, any more than he would plant Apple seed in an orchard instead of trees; and yet European vineyards are to this day planted with cuttings instead of rooted plants, just as they were two or three thousand years ago, and about the same class of implements are used in their cultivation. The cuttings should be made in autumn, and before the vines have been subjected to a very low temperature. When the vines are pruned, the canes may be cut into lengths of six to eight inches, leaving not less than two buds on each, as shown in figure 100. If the wood is short jointed, a cutting of this length will have three, and perhaps four buds upon it; if so they are all the better, as roots usually start from each bud although not always the first season. The lower end of the cutting should be cut off smooth and close up to the base of the bud, as shown, and the top an inch above the bud. When the cuttings are prepared, they may be buried in sand or moss in a cellar, or in a dry place in the open ground and below the reach of frost. In spring the cuttings

Fig. 101.　　SINGLE-BUD CUTTING.　　Fig. 102.

are taken out and planted in the same manner as other ripe wood cuttings, the upper bud being left just level with, or a half inch below the surface of the ground.

In the fall the rooted cuttings are taken up and heeled-in, and protected in winter if necessary. In warm climates, and where the ground does not freeze to any considerable depth in winter, the cuttings may be planted out in the vineyard in the fall, the long roots shortened, and the young cane cut back to within two buds of the old wood. In some localities, mulching the cutting bed will be beneficial in keeping it moist, but in others it might do more harm than good; but the cultivator of such plants must use his own judgment in such matters, as he is supposed to know something of the climate of the region wherein he is living. Nearly all of our cultivated varieties of the grape may be readily propagated by cuttings in the open ground, but an occasional one, like the Delaware and Norton's Virginia, require a little more care than others; the cuttings must be kept in a moist and rather warm place during winter in order to have the root-forming process somewhat advanced by the time they are needed for planting. What are called single-bud cuttings are made of the same kind of wood as the long cuttings, but with only one bud on each, as shown in figure 101, which represents the cutting of the usual size and length. Some propagators cut off a slice of the wood on the lower side, as shown in figure 102, thereby exposing more of the alburnum than when merely severed at the ends. But the shape of the cutting may be varied to suit the fancy of the propagator, so long as a sufficient amount of wood— but not too much—is left attached to the bud. These single bud or short cuttings are usually employed for propagating scarce varieties, and under glass during the winter months. The cuttings are planted in shallow boxes or frames filled with pure sand, and then placed in a propagating house where they can be given gentle bottom heat. These cuttings may be laid flat or thrust into the sand at a slight angle, but the bud ought not to be covered much more than a half inch when in position. Water must be applied liberally, and the temperature of the cutting bed kept at about sixty or seventy degrees until the cuttings are well furnished with roots, and the new growth from the bud is from two to three inches in height, then remove the cuttings from the sand and pot off singly in two or three inch pots. After potting, the plants may be returned to the frames, or placed in others where the air will be somewhat confined and moist for a few days, or until the plants have become established in their new position. Single-eye cuttings may be forced early in winter, but the most usual practice is to delay the operation until about the first of February in this climate. The wood, however, to be used for cuttings should be taken in early in the winter and stored in the cellar, or where it will not become dry and shrivelled. If by accident it should get very dry, the cuttings may be thrown into warm water and allowed to soak a few hours before placing them in the sand.

Cuttings of the green or unripe wood are sometimes employed in propagating rare and scarce varieties, but unless the plants are given extra care, they are seldom as strong and healthy as those raised from mature wood. The mode of operation is usually as follows: In the autumn pot the vines to be propagated or plant in a border within the propagating house, making the soil so rich that the vines will not suffer for want of nutriment. When they have made a growth of a foot or more, some of the shoots may be removed for cuttings, but do not cut back all the young growth at one time, as this would severely check the vine, but a

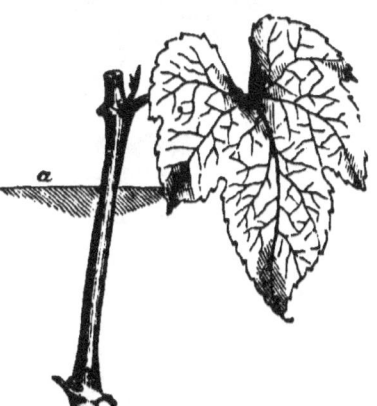

Fig. 103.—CUTTING OF GREEN WOOD.

few cuttings at a time may be taken without injury. The young shoots may then be divided into pieces of two buds each, the lower end cut off square across close to the base of a bud, and the upper leaf left entire, as shown in figure 103. The cuttings, when prepared, are planted in sand, about two-thirds of their length, covered as shown—*A* represents the surface of the sand. These green cuttings must be placed in close frames, frequently syringed overhead, and the temperature should not be allowed to fall much below sixty degrees, and if kept at eighty or ninety degrees the roots will push out all the more rapidly. These cuttings may thenceforward receive the same treatment as those of ordinary tender greenhouse plants, and in spring transferred to frames in the open ground; or, if well ripened off, set out in nursery rows, when the weather will permit, in early summer.

Layering is one of the most simple and certain modes of propagating the Grape and it is no doubt the oldest, for whenever the canes of wild vines come in contact with the earth they emit roots and thus become layers. But to facilitate the emission of roots we bend down a cane and cover that part on which we desire roots with soil, and this layer is but a cutting which is left attached to the parent plant, and derives nourishment therefrom, until it has produced roots of its own. Layers may be put down in fall or

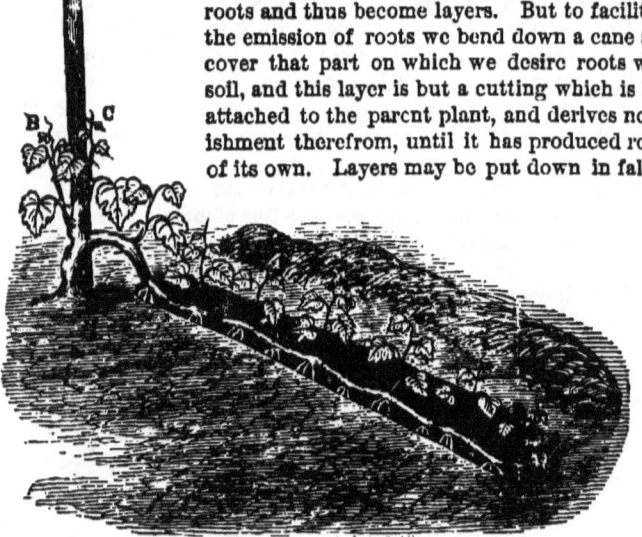

Fig. 104.—LAYERING THE GRAPE.

spring, and the young canes of the season's growth are preferable to older wood, as they produce roots the most freely and readily, no notching, twisting or tonguing being required. Vines grown expressly for layers should be planted about six feet apart, and headed back in the fall in order to force out a larger number of strong canes the following season. The growing canes may be layered in summer, but the better plan is to allow them to grow unchecked the entire season and layered the next, each cane making one strong layer after it is cut back to a convenient length. But if all the canes of a vine are layered, it would too severely check its growth, and the more usual practice is to use only one or two of the canes in any one season, and cut back the others so that new shoots will be produced for use the following year. If a larger number of layers are desired than can be secured by making one plant of each layer, then some of the strongest shoots may be laid in a shallow trench, as shown in figure 104. Select the largest cane or canes—as the case may be—for layers, and cut it back to six or seven feet, if it is longer; then cut the other canes back to within three or four buds of their base. In spring, after the buds begin to swell, layer the cane as follows: Dig a shallow trench four to six inches deep and of a sufficient length to receive the cane; now bend it down and fasten it in the trench by hooked pegs. The cane may be bent in almost any direction from the parent stock most convenient, but it should be laid flat in the bottom of the trench. The layered cane may remain in this position uncovered until the new shoots appear along its entire length, and if allowed to grow every bud will produce a shoot and a plant, but it is better to rub off at least one half of the buds and raise a less number of stronger plants. When the young shoots have grown to be five or six inches long, a little soil may be drawn in and the layered cane covered with an inch or more of soil, and a small stake placed by the side of each shoot, to which it should be tied later in the season. In the accompanying illustration the layer is shown in the trench as it appears when the young upright shoots are a foot or more in height, also the roots as they appear later. The shoots ($B, C,$) growing from the main stem may be preserved for layering the following season, or for bearing fruit. In the fall the layered cane is dug up and divided into as many plants as there are upright shoots, each with the roots at its base. The principal advantage of layering as a mode of propagation is that certain species and varieties, which are not readily propagated by cuttings, may be made to produce roots on layers.

Grafting the Grape is a very ancient mode of propagation, and is fully described by most of the old Roman authors of works on agriculture. The cions may be inserted by any and all the different methods practised in grafting woody plants—cleft, triangular, side, tongue and splice, and even by in-arching and approach. The most usual method is to insert the cion in the crown of the plant below the surface of the ground, tying it in with bass or other similar material, and then banking up with earth about the cion; no wax of any kind is used in the operation, the earth being sufficient to exclude air and prevent drying. The proper time or season for grafting is still—as it has been for thousands of years—a moot point among vineyardists. The old Romans could not agree as to the best time for grafting the Grape in the vineyards about Rome; for while Julius Atticus said that the time for grafting was from the first of November until the first of June, Columella objected to this prolonged season and thought the better time was in spring after the cold weather is past. The same difference of opinion exists among vineyardists at this day, probably because climate as well as experience differ. My own experience is in favor of early grafting either in the fall or winter, then protecting the cions from frost either by covering with an inverted flower pot and straw, as I described many years ago in the "Grape Culturist," or by merely covering with a few forkfuls of coarse stable manure. In warm climates no such protection is required, and a cion set early in the winter becomes fully united to the stock before a rapid flow of sap begins in spring, and this no doubt is why Julius Atticus recommended early grafting of the Grape in his time, or nearly two thousand years ago. But early spring grafting is most generally practised in cold climates, the cions being stored in a cool place where they will remain dormant until wanted for use. The earlier they are inserted the more likely they are to unite, and if not inserted early it is better to wait until the leaves have unfolded on the stock and the sap has become somewhat thickened through evaporation from the leaves. When vines are grafted on roots below the surface, the cion, if not prevented, will throw out roots of its own, and the benefit of a strong, old stock will be only temporary; but if these surface roots are removed once or twice a year the old stock may continue to exercise its influence upon the cion for many a year. Grafting by approach above ground is readily accomplished by the usual mode during the early summer months, and this is probably preferable to grafting below the surface where it is necessary to use what are called "phylloxera resisting stocks," as some of our native species of the vine are called in France and California.

Weigela.—See *Diervilla.*

Wistaria.—Very rapid-growing, woody, climbing plants, with pea-shaped flowers in long, drooping clusters. One species (*W. frutescens*), native of the United States, and one (*W. Sinensis*), in China. There are several varieties of both species in cultivation, nearly all now common in our gardens. Propagated by seeds, which are produced in great abundance on old plants, by layers and grafting. Cuttings do not usually strike root very readily when planted in the open ground, but will succeed if planted in sand under glass where they will receive moderate heat.

Zanthoxylum (Prickly Ash).—A small genus of ornamental and useful shrubs; the bark, leaves and fruit extremely pungent and aromatic. The common native species found in our Northern woods is known as "Toothache tree." Another species is found in the Southern States, and several in the East and West Indies and other tropical countries, besides in China and Japan. One species from the latter country (*Z. piperitum*) is nearly hardy in my grounds, the terminal shoots only being injured in winter. Propagated the most readily by cuttings of the roots.

CHAPTER XX.

HERBS, TUBERS AND BULBS.

In the following pages I purpose giving only very brief hints in regard to the propagation of herbaceous, bulbous, tuberous, and some suffrutescent perennial plants omitted in the preceding chapters. Although the principles governing their growth and propagation are the same as with other kinds, still it may often occur that a hint in relation to some simple mode of increasing a species or variety is of more value to the inexperienced than a learned treatise on the subject. Certain modes of propagation well known to one person may not be to another; consequently, in attempting to impart information in a work like this, the author is obliged to presume somewhat upon the inexperience of his readers. As a rule, it may be said that all kinds of herbaceous plants, such as Carnations, Phloxes, Petunias, Verbenas, Snap-dragons, and all similar kinds having stems bearing leaves, may be more or less readily propagated by cuttings of the tender or half-ripened shoots, placed in frames or under a bell glass, where the air will be somewhat confined and moist, while at the same time a moderately high temperature can be secured. In temperate climates this mode of propagation may be practised with success in ordinary hot-bed frames, without bottom or artificial heat, during the summer months.

The soft-wooded greenhouse plants, such as Geraniums, Fuchsias, Lophospermums, Begonias, etc., may be increased under the same conditions, as well as many of the more succulent kinds, like the Ageratums, Alternantheras, Alyssums and Coleuses; but a propagating house, built especially for such purposes, is always preferable to cheaper structures of this kind, because of the facilities afforded for controlling the temperature in our changeable climate. In making cuttings of soft-wooded and herbaceous plants, it is always best to cut through close under a joint or bud, although there are kinds which strike root so readily that it will make very little or no difference where or how the stems are divided. The leaves on the lower part of the cutting—that part to be buried in the sand or soil—must be removed, else they are likely to decay and increase the danger of what is termed "damping off." It may also be advisable in some cases to remove a part of each leaf on the cutting, especially if the leaves are large and soft; but with most plants propagated by cuttings of the young shoots, the terminal leaves may be left intact.

In the application of water to such cuttings, the propagator must ever depend upon his own observations and judgment. The cuttings must not be allowed to flag for want of moisture, neither should the atmosphere in the frames be kept constantly saturated. Ventilation must also be attended to, and more air admitted as the cuttings advance in growth and in the production of roots, than when first placed in the frames or under bell glasses. Cuttings of the hardy and half-hardy herbaceous plants, such as Carnations, Phloxes and Hollyhocks, do not require so high a temperature to insure the production of roots as those of the Coleus, Acanthus, Achimenes, Begonias and other kinds, natives of tropical climates. There are usually more cuttings lost by attempting to force their growth by a high temperature than in keeping it too low.

In propagating bulbous, tuberous, and other plants with large, fleshy roots, it should be kept in mind that no great amount of moisture is required until the leaves have been produced and growth has fairly begun. They all have their seasons of growth and of rest, and this natural habit—as we may term it—has been acquired through the influence of the climate of the region in which they have lived for an unknown number of centuries. Some kinds of plants seem to submit more readily to artificial conditions than others, but the most satisfactory results will usually be secured by keeping very near to nature in dealing with the plants of any country or clime.

Acanthaceæ (Acanthus Family).—Mostly tropical herbs; a few climbers, as in *Thunbergia*, but the most highly prized belong to the genus *Acanthus*, these being large, stately, ornamental perennials, much admired for their beautiful foliage. Very useful plants for bedding out in summer. Propagated by seeds sown under glass, or by division of the roots while the plants are in a semi-dormant condition in fall or winter.

Amarantaceæ (Amaranth Family).—A large family; mostly low annual herbs; a few shrubs, and only a few genera considered worthy of cultivation, and among these are the well known *Amarantus*, *Celosia*, *Gomphrena*, *Alternanthera*, and *Iresine* or *Achyranthes*. The species of the two last named genera are perennials, and of which there are many varieties in cultivation, valued for their handsomely colored foliage. To insure a perpetuation of the bright color and variegation of the leaves, the plants should be propagated by cuttings taken from stock plants kept over for the purpose. Cuttings taken off in March and April will usually become sufficiently strong and well supplied with roots for planting out later in spring.

Amaryllidaceæ (Amaryllis Family).—A very large family of elegant ornamental plants, mostly bulbous, but a few, such as the *Agave* (American Aloe) have stems and large fleshy roots. The most familiar genera are the *Amaryllis*, *Crinum*, *Pancratium*, *Narcissus*, *Galanthus* (Snowdrop), *Hippeastrum*, and *Hæmanthus*. A few, such as the *Narcissus* and *Galanthus*, are hardy, but they are mainly greenhouse plants. New varieties are raised from seeds, and as these are rather fleshy they should not be covered very deep, especially if soil is used for this purpose. I have had excellent success by scattering the seed over the surface of leaf mold, and then spreading over them a few shreds of moss, covering all with a bell glass or a pane of window glass laid flat on the top of the seed-pan or pot. The young plants may be potted off as soon as they are large enough to be readily handled. All of the different genera require a deep rich soil with good drainage, whether cultivated in pots or in the garden. Varieties are propagated by offsets, which are usually produced in great abundance, although in a few kinds the old bulbs produce buds rather slowly and sparingly.

Apocynaceæ (Dogbane Family).—A family composed of trees, erect and twining shrubs, and many low herbs, mostly containing an acrid, poisonous juice. The common Oleander (*Nerium*), is a well-known evergreen shrub belonging to this family. Among the low herbaceous and evergreen genera, the *Apocynum* (Indian Hemp), *Amsonia*, and *Vinca* (Periwinkle), are common border ornamental plants. Propagated by cuttings and divisions of the clumps or stools in spring.

Aroideæ or *Araceæ* (Arum Family).—A large order of herbaceous perennial plants with tuberous rhizomes. The most familiar genera in cultivation are *Alocasia*, *Amorphophallus*, *Anthurium; Caladium*, *Colocasia* (Tanya), *Diffenbachia*, and *Richardia* (Calla). The flowers are very minute, unisexual or perfect, produced on a central organ called a spadix, and this surrounded by a large spathe, which is sometimes—as in the *Amorphaphallus Rivieri*—two feet or more in diameter, and emiting a

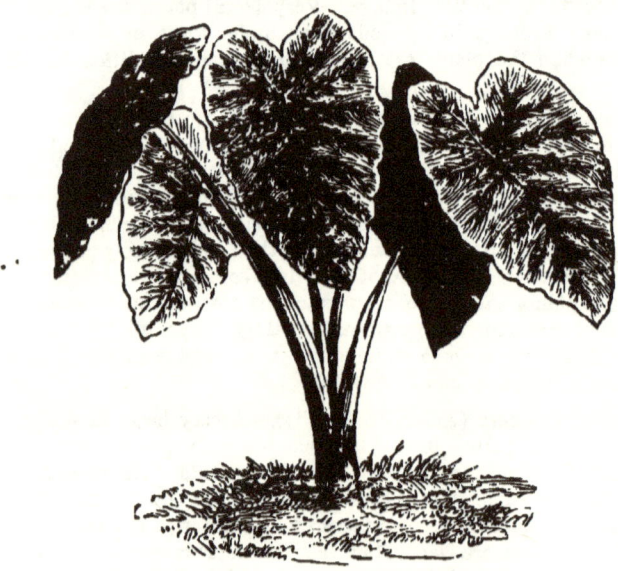

Fig. 105.—COLOCASIA ESCULENTA.

most disagreeable odor. The rhizomes of some of the genera contain an acrid watery juice, as in the wild Indian Turnip (*Arisæma*), while in others, such as the *Colocasia esculenta*, they are edible when cooked. This edible species is known in our Southern States, where it has become naturalized, as the Tanya. It is largely employed as a bedding-out plant in the Northern States, the leaves growing to an immense size, and the general habit and form of the plant is shown in figure 105. The *Richardia* or Calla Lily, as it is called, is a well-known and common window and greenhouse plant. The *Caladiums* are what may be termed hot-house plants, requiring a high temperature and moist atmosphere to insure the full development of their brilliant colored and handsomely variegated leaves. The *Diffenbachias* are of a more stocky growth than the *Caladiums*, the large leaves springing from a central fleshy stem, which in some species is six to eight feet long. This genus possesses a very poisonous acrid juice, and the propagator should bear this in mind when dividing the plants or removing the leaves. The *Anthuriums* are perhaps the most showy of the family, on account of the immense size of the bract-like spathe, which bend backward instead of folding around the spadix, as in the common Calla and others. All members of this family are moisture-loving plants, and need a copious supply, especially when growing rapidly. Propagated by seed, and by divisions of the corms, or the naturally produced offsets. Some of the genera throw off small tubers in large numbers, others few; but on all old and strong tubers, small buds may be found which may be cut out when the plants are at rest, and if placed in gentle heat will usually produce roots from around the crown or eyes. Seeds of the Calla will usually produce blooming plants under favorable conditions in a twelvemonth, but for those of some of the other genera, two to three years are required from seed to obtain even plants of moderate size.

Begoniaceæ (Begonia Family).—A small family, and only one genus in cultivation, and that the Begonia; but of this there are at least three hundred species and an innumerable number of garden hybrids and varieties. The Begonias are mostly succulent herbaceous or somewhat woody-stemmed plants, with unequal-sided leaves, which no doubt suggested the common name of "Elephant's Ear." In some of the species the roots are very thick and fleshy, in others, distinctly tuberous. The flowers are usually showy, sometimes of enormous size, and of various colors—white, rose, scarlet, yellow, and all intermediate shades. Seeds minute, but produced in great abundance. One species, the *B. socotrana*, has an herbaceous annual stem, at the base of which small bulblets are produced, and these should be kept dry during the summer months and forced into growth during the winter. All the Begonias are of easy culture, thriving in a warm greenhouse. Readily and rapidly propagated by seed, and by cuttings of the stems and leaves. The minute seeds should be sown on the surface of light soil or pulverized charcoal and moss, and not covered with soil, but the boxes or pans covered with a pane of window glass or a common bell glass. Water should be given freely and often through a fine rose syringe or atomizer, and the temperature kept at about sixty degrees, or slightly above.

Boraginaceæ (Borage Family).—Mostly rough, hairy, annual and perennial herbaceous plants. Some, like the common Comfrey (*Symphytum*), with thick fleshy roots containing a mucilaginous juice. The genera most highly valued are *Anchusa*, *Mertensia* (Lungwort), *Heliotropium* (Heliotrope), *Myosotis* (Forget-me-not), and *Symphytum* (Comfrey). The different species and varieties of the *Anchusa* and *Symphytum* are usually propagated by division of the clumps or cuttings of the roots, and the others by seeds or cuttings of the tender shoots.

Bromeliaceæ (Pine-Apple Family).—The most familiar genera of this Family are *Ananassa* (Pine-apple), *Æchmea*, *Bilbergia*, *Bromelia*, and *Tillandsia*, one of the species of the latter being the well-known Spanish moss of Florida and other parts of the South. The *Ananassa* or Pine-apple is the most important genus in this family, and now exists in both its wild and cultivated state in all the warmer or tropical countries of America. There are a number of varieties in cultivation, and all are propagated by cuttings made of the sprouts which naturally spring up around the base of the main stem. When the fruit is cut off the sprouts appear in more or less abundance, and these are slipped off and planted in sand where they can be given bottom heat. The *Æchmeas*, *Bilbergias*, and *Bromelias* are readily propagated in the same way, all requiring a rather high temperature and abundance of water to insure vigorous growth and free blooming plants. When at rest they need but just sufficient moisture to prevent shrivelling.

Cactaceæ (Cactus Family).—An immense order or family of succulent or fleshy plants, mostly destitute of true leaves, the functions of these useful organs on other plants being performed by the green rind of the columnar, flattened, or other shaped stems and branches. These plants are most abundant in the warmer regions of North and South America, but some of the species grow at high altitudes in the tropics, and others extend far northward, where they are subjected to a temperature many degrees below zero in winter. The most popular and best known genera are *Cereus*, *Mammilaria*, *Melocactus*, *Epiphyllum*, *Echinocactus*, *Phyllocactus*, *Opuntia*, and *Pereskia*. We have no family of plants that thrive under neglect nor respond more fully to good care than the Cactuses. They all need moisture while growing, and very little or none at all white at rest. There are many species found in the higher regions of the West and South that withstand a temperature of ten or more degrees below zero in their native habitats, where rains seldom fall in winter, and yet these same species are quickly destroyed by slight freezing in a moist climate or atmosphere. Cactuses from the cool and elevated regions of New Mexico, Arizona and Old Mexico should not be exposed to the direct rays of the sun when cultivated in our Atlantic States, as the heat is often much greater than it is in their native habitats, and shade during the middle of the day should always be given to plants placed outdoors during the summer months. Propagation is effected by seeds, cuttings, and grafting. Seeds are rarely employed except for producing new varieties, but cuttings of the species and varieties with columnar and branching stems strike root very readily if placed in sand or any light, loose soil. The cuttings should be watered very sparingly until they are well supplied with roots. The small globular Melon and Hedgehog Cactuses are usually propagated by removing the small sprouts appearing at the base or sides of the old plants. Grafting the smaller species, such as the Crab Cactuses (*Epiphyllum*), on the stronger, upright-growing species, is extensively practised by florists, and with great success. The stocks usually recommended are the *Pereskia* and *Cereus* for the *Epiphyllums*, and the smaller, globular-shaped genera. For many years I have employed the *Cereus speciossimus* as a stock for the varieties of Crab Cactus (*Epiphyllum truncatum*), and the plants appear to thrive as well as on other and stronger-growing stocks. In grafting, all that is necessary is to place the fresh cut surface of the cion against a similar surface of the stock, and keep the two in contact until a union is formed. When small species, like the Crab Cactuses, are grafted, it is best to use a wedge-shaped cion and insert it into a cleft at the top of the stock, then thrust a sliver of pine wood through both cion and stock, or in other words, pin them together. A sharp, strong spine of a cactus will answer for a pin, and this may be left in place until a firm union is formed, and then withdrawn with a pair of nippers. In grafting the small globular-shaped species on stocks of the tall-growing kind, scoop out a little of the pulpy matter from the top of the stock, then set the freshly cut cion in this depression and tie it firmly in place. The soil for Cactuses should be made up of rather coarse materials, such as partly decomposed sods, with plenty of drainage when grown in pots.

Campanulaceæ (Bellflower Family).—A family of herbs and sub-shrubs. The genus most highly prized and best known is the *Campanula* or Bellflower. Flowers mostly blue or white, with intermediate shades. There are annual, biennial, and perennial species, the latter being most extensively cultivated, as they are nearly all quite hardy and thrive in any good garden soil. Readily propagated by seed sown in frames, or in the open ground in spring, and by divisions of the roots.

Capparidaceæ (Caper Family).—Herbs and shrubs, rarely trees, distributed throughout the warmer regions of both hemispheres. There are twenty-three genera, and nearly or quite three hundred species. The best known genera are *Capparis*, *Cleome*, and *Cratæva*. The latter yields the curious Garlic Pear (*C. Tapia*), of Central America, a tree growing to the height of thirty to forty feet. *Capparis spinosa*, native of the Levant, yields the well-known Caper of commerce, while several species of *Cleome* are cultivated in greenhouses for their flowers. All readily propagated by seed, or cuttings of the young, tender shoots planted in sand under glass.

Caryophyllaceæ (Pink Family).—A very large family of low-growing herbs, consisting of annuals, biennials, and perennials. The most interesting genera are *Dianthus*, *Lychnis*, *Silene*, and *Cerastium*. The common Carnations (*Dianthus Caryophyllus*), especially the monthly or perpetual bloomers, have become exceedingly popular of late years and are now extensively cultivated by florists, as they bloom most freely during the winter months. They succeed best in a cool greenhouse, or pits where the temperature can be kept at about sixty degrees during the day, and not much below forty-five at night. The plants are almost hardy, but require a moderate heat to develop the flower buds. Propagated by layers and cuttings, and new varieties are raised from seed. Cuttings strike root so freely that this is the usual mode of propagation.

They should be taken off late in winter or early spring and planted in pure sand under glass, but they do not require a high temperature or a very copious supply of water while the roots are being produced. A temperature of sixty degrees will insure the production of roots, and with less danger of the cuttings damping off than if exposed to a higher temperature. The cuttings should be from two to three inches long, the base cut just below a joint, and the leaves from the lower part removed, while those above are shortened to about one-half their original length. Carnation cuttings will usually strike root very readily in an ordinary greenhouse, without placing them in close frames or where they will receive bottom heat. It is well to shade the cuttings, or protect them from the direct rays of the sun, for a few days after planting. A few species of *Cerastium* (Mouse-ear Chickweed) are cultivated in greenhouses, and others for edgings of beds in summer or as border plants. The same may be said of *Lychnis* and *Silenes*, and all are readily propagated by seeds, cuttings, or division of the roots.

Cistaceæ (Cistus or Rock Rose Family).—A small order of elegant shrubs or sub-shrubs, with very showy flowers of various colors, from pure white to purple and yellow. The best known genera are *Cistus* and *Helianthemum*. There are many species and varieties of *Cistus* in cultivation, some of them quite hardy in our Northern States; others are tender, requiring the temperature of a cool greenhouse in winter. The flowers are very handsome, but seldom last more than one day; consequently are of little value for cutting or using in bouquets. The *Helianthemums* are very similar to the Rock Rose in general appearance, but not usually of as strong and robust habit. Some of the species are annuals, but there are many half shrubby perennials. Propagated by seeds, division of the clumps, and by green cuttings planted under glass and treated as usual with such cuttings.

Commelinaceæ (Spiderwort Family).—A large and widely distributed family of herbaceous plants, mostly tropical. Only a few genera in cultivation, and the two most deserving attention are *Commelina* and *Tradescantia*. The latter is usually represented in gardens by the very common Spiderwort (*T. Virginica*), and in greenhouses by several varieties of the Striped-leaved Spiderwort (*T. zebrina*). The *Commelinas* are only occasionally cultivated; a few species are grown in greenhouses for bedding out in summer, and among these there are several tuberous-rooted kinds, which should be lifted in the fall and stored in a dry, warm place until spring. All readily propagated by seeds and divisions, and the trailing kinds by cuttings and layers.

Compositæ (Composite Family).—This is the most extensive family of the entire vegetable kingdom, containing between seven and eight hundred genera, and fully ten thousand species. They are mostly herbs, but a few being shrubs; the flowers, collected in a head on a common receptacle, usually surrounded by an involucre bract, as seen in the common Sunflower, Artichoke and single Zinnia. The genera alone are far too numerous to name here, and I can only mention a few of the most important and valuable among the herbaceous perennials. Aster—only a few perennial species are cultivated. The popular annual China Aster is a *Callistephus*. *Artemisia*—only one species is of special value, and that the *A. Absinthium* (Wormwood), a low-growing, hardy plant, possessing some medicinal properties, and another is largely used for flavoring the French liqueur known as absinthe. *Anthemis nobilis* is the well-known garden herb, Chamomile. *Achillea*, or Sneezeworts, are mostly low perennial weeds, but a few are handsome border plants, and *A. Ptarmica pleno* has pretty double white flowers. *Antennarias* are known as Everlasting, and one—the Pearly Everlasting—is extensively cultivated in Europe, and the flowers dried for winter bouquets. *Liatris*, of which there are many hardy indigenous species, are best known in cultivation under the name of Blazing Star. They are tall-growing plants, with rather thick and woody corms or tubers at the base of the stems. *Arnica* and *Inula* (Elecampane), are genera yielding medicinal properties more or less in repute. *Gaillardia* and *Gazania* are showy greenhouse plants, also employed for bedding out in summer. These plants are raised from seeds, or by divisions of the roots and cuttings of the young shoots. All of the hardy genera are most readily propagated by division. *Chrysanthemum* has of late years become one of the most popular genera in the family, especially the Chinese species (*C. Indicum*). Propagation may be effected by seeds, cuttings, divisions or suckers. Varieties can only by perpetuated by the last three modes. Cuttings of the young shoots strike root very readily under glass and with moderate heat, and the plants require only good, rich soil, plenty of moisture, and plenty of room in which to expand.

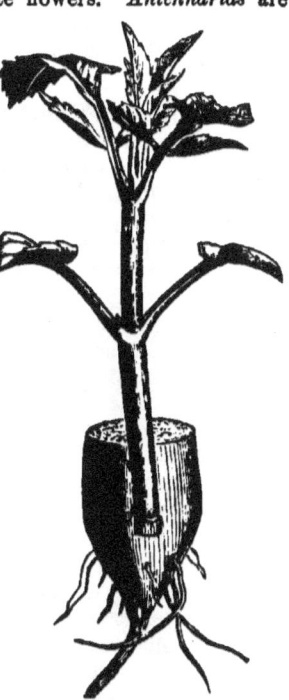

Fig. 106.
GRAFTING THE DAHLIA.

There are now more than a thousand named varieties of Chinese and Japanese Chrysanthemums in cultivation, and scores of new ones are brought forward every season. The Dahlia is another very popular genus of the Composite Family of plants. There are but few distinct species, but an immense number of cultivated varieties. Flowers usually very large and showy, and of many shades of color; roots tuberous, several tubers usually attached to the base of the stem, and the eyes or buds always at the apex of each tuber, or on the stem. The most usual mode of propagation is by dividing the clump of tubers, care being taken to preserve at least one eye or bud on each. This dividing is done in spring, the tubers having been kept in a warm, dry place over winter. To multiply scarce varieties, the tubers may be potted, and as the sprouts push they may be slipped off and treated as cuttings, planting in sand or light soil, where they will receive a little bottom heat. When roots have formed in the cuttings they should be placed singly in small pots. Cuttings made of either the side or terminal shoots, taken from large plants, strike root very readily under glass. Grafting is sometimes practised, a small piece of a tuber answering for a stock, the cion being inserted as shown in figure 106, and held in place with a ligature of bass or fine twine. The grafted tuber is then planted in a frame and given the same care and attention as usually bestowed upon cuttings.

Convolvulaceæ (Morning Glory or Convolvulus Family).—A family of half a dozen or more genera, but nearly seven hundred species. The common Morning Glory (*Ipomœa*), and *Quamoclit* (Cypress Vine), are well-known annual climbers of this family. A few genera, such as the *Cuscuta* (Dodder), and *Calystegia* (Bindweed), are weeds which have become intolerable nuisances in many localities. The Sweet Potato (*Ipomœa batatas*), is the most valuable member of the family, although there are many other species of the same genus that are extensively cultivated for ornamental purposes. The perennial, tuberous-rooted species are propagated from sprouts, which are forced out in great abundance by placing the tubers in a hot-bed, or where they will receive gentle bottom heat. The sprouts, when a few inches long, are pulled off and planted out separately. This is the usual mode of propagating the varieties of the Sweet Potatoes, but this and other perennial species may be multiplied by cuttings of the vines, taken off at almost any time during the summer.

Crassulaceæ (Orpine Family).—A large family composed principally of succulent herbs. There is about a dozen genera and several hundred species. The best known are *Bryophyllum, Crassula, Sedum, Rochea* and *Sempervivum*. The common House Leek (*Sempervivum tectorum*) is a familiar and good representative of the family, and while the other genera may not be as hardy or as difficult to kill out, still they may be readily propagated by divisions, or cuttings of leaves and stems.

Cruciferea (Mustard Family).—A large order containing many kinds of useful and edible, as well as ornamental plants. There are 170 genera and nearly 1,200 species. Many of our common garden vegetables, such as Cabbage, Cauliflower, Cress, Horseradish, Mustard and Turnip, belong to this family, but the propagation of these plants is too familiar to all cultivators to be repeated here. Among the ornamental genera, the following are worthy of some attention: *Arabis* (Rock Cress), *Alyssum* (Sweet Alyssum), *Iberis* (Candytuft), *Chieranthus* (Wallflower), *Hesperis* (Garden Rocket), and *Matthiola* (Stock or Gilliflower). These, however, are all easily propagated by seeds, divisions, or from cuttings of the young shoots, planted in an ordinary greenhouse, or under a frame during the summer months.

Curcubitaceæ (Gourd or Cucumber Family).—A family of succulent, climbing or trailing plants, furnished with solitary lateral tendrils. They are mostly annuals, and many of the genera yield large and delicious fruit, such as the Melons; others are cultivated for ornament, such as the small Gourds and the *Bryonias*; the latter are greenhouse perennials with tuberous roots. Propagated by seed, layers, and the tuberous rooted species by division of the tubers when in a dormant state.

Cycadaceæ (Cycas Family).—A family of small, Palm-like plants, closely related to the conifers, the male flowers being in cones, and the female consisting only of ovules on the edges of what may be termed abnormal leaves. There are nine genera, all tropical or sub-tropical. The Sago Palm (*Cycas revoluta*) and the Coontie (*Zamia integrifolia*), of Florida, are familiar representatives of this family. Propagated by seed, but mostly by suckers or sprouts that spring up about the stems of old plants.

Cyperaceæ (Sedge Family).—A large family of low, grass-like, mostly perennial plants, with minute flowers in spikes or heads. Only a few of the genera are of horticultural value, and among these *Carex*, *Cyperus*, *Papyrus* and *Scirpus* are the best known. The *Cyperus esculentus* is the well-known Chufa, the small tubers of which are quite sweet and edible, sometimes cultivated for feeding swine and sheep. The Variegated Rush (*Cyperus alternifolius variegatus*) is a handsome perennial greenhouse plant. *Papyrus antiquorum* is supposed to be the Bulrush of the Nile, from which paper was first made. It is in common cultivation in greenhouses. All the genera require an abundance of moisture. Propagated by division of the roots or tubers, and by seeds.

Dioscoreaceæ (Yam Family).—A family of twining herbs and shrubs, mostly natives of tropical climates, and the larger proportion producing tubers, used as food the same as the common potato. *Dioscorea sativa*, is the common yam of the West Indies, and *D. batatas* is the Chinese yam or potato, introduced into this country about thirty years ago, and highly extolled as a tuber likely to supersede the common potato; but while the tubers are of excellent quality, they penetrate the earth so deeply that it costs more to dig them than they are worth. This species is now cultivated as a hardy climbing ornamental vine, its thick leaves making it an excellent climber, while its dull-looking flowers are so fragrant that it is offered by some as the "Cinnamon vine." All readily propagated by dividing the roots or tubers, and by cuttings of the stems. Some of the species produce small aerial tubers in the axils of the leaves, and these are utilized in their propagation.

Droseraceæ (Sundew Family).—A small family of low annual and perennial bog-herbs, with flowers consisting of from four to eight persistent sepals, and a similar number of petals. Leaves variable, often bristly fringed. There are six genera, and about a hundred species in this family. The best known genera are *Dionæa*, *Drosera* and *Drosophyllum*. The *Dionæa muscipula* is a native of the low peat bogs near Wilmington, N. C., and has long been known under the common name of "Venus's Fly-trap," from the way the leaves close up and catch flies and insects which may alight on the inner surface of the expanded leaves. Only one species of this genera is known, but of the *Drosera* there are about ten known, one half native of our Southern States, and the other of Australia. They possess the same irritability as the *Dionæa*, catching small insects. The *Drosophyllum lusitanicum* is a greenhouse shrub, native of Southern Europe and Africa, but its leaves and hairs are not sensitive to the touch. It is propagated by cuttings, but the species of the two first named genera may be propagated by seed or by divisions. They thrive best in light, peaty soils, or in beds composed principally of Sphagnum moss from the swamps.

Filices (Fern Family).—An immense order, of about seventy-five genera, and nearly, or quite, 2,500 species. They are mostly perennial herbs, with creeping or ascending root-stocks, a few climbing, others shrubby or arborescent. Ferns are most abundant in warm, moist climates, but there are many low growing species, extending far into the colder regions of nearly all parts of the world. The leaves, or fronds as they are usually termed, are tufted or alternate on the root-stocks; some are simple, but the larger proportion are compound and variously divided; and the segments, pinnæ and pinnules are widely variable in numbers, size and forms. Ferns are called flowerless plants, because the organs of fructification are microscopic, and the spores, which answer the same purpose as seeds in the higher order of plants, are usually collected in masses, or a *sorus*, on the under side of the fronds, as in the common wild Adder's Tongue ferns (*Ophioglossums* and *Botrychiums*). The propagation of ferns is usually affected by varying the mode in accord with the different habits of the numerous sub-families and genera. The most common one is by divisions of the creeping rhizomes, or by the little bulbils forming naturally on the fronds of some of the genera, such as in the *Aspleniums*. The best time to divide the rhizomes, or those kinds growing in small slender clumps with thread-like roots, is just before growth commences, and while the plants are in a semi-dormant state. Taken as a family, the ferns need a great amount of water both over-head and at their roots, but what is usually called stagnant moisture at the roots is highly injurious, consequently good free drainage is of the utmost importance. Propagation by spores is always an interesting mode, even when not necessary for the purpose of obtaining a stock of plants, because of the chances afforded of producing new varieties; and when a number of different species are grown together in the same house, variations from normal types may be more pronounced than where only a few or a single species is cultivated. In preparing seed pans, or boxes, for the reception of the spores, cover the bottom with broken pots or pieces of brick,

HERBS, TUBERS AND BULBS. 321

and over these place old, half-decomposed sod, then fill up with a mixture of fine leaf-mold and silver sand. Smooth the surface, and make it as level as possible, then scatter the spores and leave them without any covering of soil or other material. Set a bell glass over the seed pan, or, if deep enough, a pane of window glass laid flat on the top of the pan or box will answer equally as well. Water must be applied by placing the bottom of the seed pan in a shallow saucer or other vessel, leaving it in this position until the water rises to the surface of the soil within; then remove it. Watering overhead is not practicable until the plants appear, except it be applied with an atomizer. The frond from which the spores are to be obtained should be cut when the *sori* begin to turn brown, and by passing the finger over them a few can be easily rubbed off. Lay away the frond in the shade for two or three days, then scatter the spores in the seed pan by holding the frond over it, and snapping the back with thumb and finger. When the young plants appear, and are large enough to be removed safely, they should be pricked off in small clumps, and lifted out on the point of a knife. When of the size shown in figure 107 they may be potted off separately, using small thumb pots for the purpose. It is well to keep them in the house, where they will be shaded and in a close atmosphere, until they are well established.

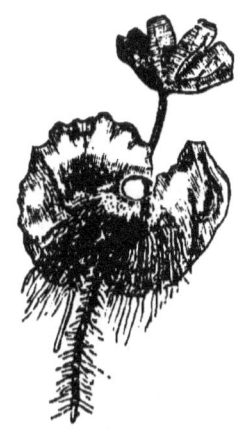

Fig. 107.—FERN SEEDLING.

Fumariaceæ (Fumitory Family).—A small order of herbaceous plants, closely related to the Poppy Family (*Papaveraceæ*), and by some botanists considered as only a tribe of the latter. The most interesting genus in the family is *Dicentra*. This genus contains several hardy perennial species of ornamental plants, a few indigenous to the United States, such as *D. cucullaria* (Dutchman's Breeches) and *D. Canadensis* (Squirrel-corn); but the most showy and valuable species is the *D. spectabilis* (Bleeding Heart), brought from Northern China. This is truly one of the most graceful as well as beautiful hardy herbaceous

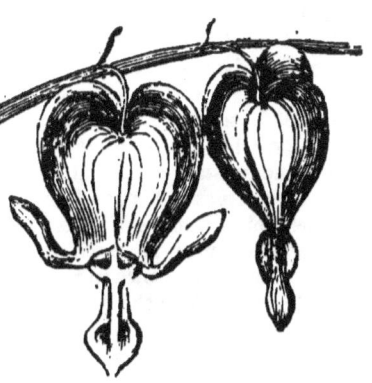

Fig. 108.—FLOWERS OF DICENTRA SPECTABILIS.

plants in cultivation, and a universal favorite. It is readily propagated by division of the large clumps of fleshy roots, or by cuttings of the succulent stems, planted either in frames, or in a shady place in the open ground. Cuttings of the blooming stems, taken just as the flowers begin to drop off, root quite readily in the open air, if water is applied freely during dry weather. Of the other genera very few are cultivated, although some of the perennial *Corydalis* are occasionally seen in botanical collections and private gardens. All readily propagated by seed and cuttings of the succulent stems.

Gentianaceæ (Gentian Family).—Mostly smooth annual, biennial or perennial herbs, a colorless bitter juice being a characteristic of the entire order. Of the forty-nine genera of this family, a very small number are represented in gardens and greenhouses. The most popular of all are the true Gentians (*Gentiana*), and some of the very best of these resist all attempts to domesticate them. The best mode of propagation is by seed, and this had better be scattered in what may be termed a "wild garden," and the plants left undisturbed. The perennial species, however, may be raised in seed pans or frames, and the plants transplanted when large enough for handling; but the slow growth of seedling Gentians will try the patience of the most persevering of propagators.

Geraniaceæ (Geranium Family).—A large order of about twenty genera and over seven hundred species, widely distributed throughout the temperate and semi-tropical regions of the whole world. The most familiar genera are the *Geranium*, *Pelargonium*, *Tropæolum* and *Oxalis*. The latter, however, belong to a distinct tribe—*Oxalideæ*—of the order *Geraniaceæ*. The true Geraniums are rather sparingly represented among cultivated plants; the most popular species and varieties that are commonly called geraniums are really Pelargoniums, such as the Zonale, Rose-scented, Nutmeg, Oak-leaved, Ivy-leaved, Scarlets and Tricolors; but as the species have become so intermingled by hybridizing and crossing under cultivation, it is now very difficult to determine the true parents of any of the older varieties. The Pelargoniums are all perennials, and more or less shrubby. New varieties are raised from seed, which may be sown as soon as ripe, or preserved for several months and then sown in rather light soil and covered with fine earth sifted over them to the depth of not more than one-eighth of an inch. The seeds usually germinate readily in a temperature of 60 to 70 degrees, and as soon as the plants have produced a second or third pair of leaves they should be removed from the seed pans and potted separately. With few exceptions, all the varieties in cultivation are readily propagated by cuttings, made of the young succulent shoots, planted in sand in an ordinary cool greenhouse; a very high temperature is not desirable for cuttings of Pelargoniums. One or two good leaves should be left on the cutting; but with very scarce and rare varieties single-eye cuttings may be used, making these of rather firm and mature shoots. Root cuttings may also be utilized in multiplying choice varieties, taking the roots of large and rather mature plants for this purpose. Cuttings of either the shoots or roots may be made at any or all seasons, if given the protection of a house or of an ordinary garden frame covered with glazed sash. Of the genus *Oxalis*, there are are a large number of species, some natives of cold climates and, of course, quite hardy, as with our common Wood Sorrel (*O. Acetosella*); others are tender and cultivated in greenhouses. There are several bulbous or tuberous-rooted species; and at least two, found in South Ameri.a, are cultivated for their edible tubers. All are readily propagated by seed, divisions of the roots and tubers.

Gesneriaceæ (Gesneria Family).—A large family of tropical herbs and shrubs, rarely trees; often growing from scaly tubers. Flowers with somewhat irregular corollas; the corolla tube often elongated or tube-shaped. Mostly very showy conservatory plants, requiring a high temperature and a moist atmosphere during their season of most rapid growth. The most familiar genera are: *Achimenes*, *Gesneria* and *Gloxinia*. The seeds of all these genera are very minute, and should be sown on the surface of very light soil, then covered only with a bell glass and water applied with an atomizer, or through a fine rose on a watering pot or syringe. The seed requires a high temperature and a constant supply of moisture; and the young plants need to be handled with great care in transplanting, and kept shaded and moist until well established. Varieties of *Achimenes* may be readily propagated by means of the small corm produced at the base of their stems, or by cuttings of any portion of their stems, planted in light soil or in sand, then giving plenty of moisture overhead, and a high temperature. The larger number of the species of *Gesneria* in cultivation are tuberous-rooted and herbaceous, and they are propagated by dividing the tubers and by cuttings of the stems, taken when somewhat mature or hardened off after blooming. *Gloxinias* are low—almost, or quite stemless—herbs, with very large soft leaves and showy flowers of many colors. Seedlings bloom the first year, and varieties are readily multiplied by cuttings of the leaves, or from the young sprouts as they push from the old tubers in spring.

Gramineæ (Grass Family).—An immense order, and a large number of the genera of the greatest importance to man, not only yielding the bread materials of the world, but supplying the most valuable of our domesticated animals with food. The cerealia, Wheat, Rye, Oats, Rice, Sorghum and Indian Corn, are annual grasses, while the Tropical Sugar Cane, the giant Bamboo of Asia, and many of the larger ornamental species are perennials. The cultivation and propagation of all the more useful genera are so well known that they may be omitted here. It may not, however, be generally known among cultivators that the *Arundo*, *Bambusa*, *Dendrocalamus*, and other closely allied genera, may be almost or quite as readily propagated by cuttings of the stems as the *Saccharum officinarum*, or Tropical Sugar Cane. This mode is preferable to the usual one of dividing the roots for perpetuating grasses with variegated foliage.

Iridaceæ (Iris Family).—A moderately large order of monocotyledonous perennial herbs, the stems and leaves rising from somewhat fleshy root stocks, bulbs or tubers. Flowers showy, and of many forms and colors, in terminal spikes, corymbs, or loose panicles. This family is extensively represented in gardens by such genera as *Crocus, Gladiolus, Iris, Ixia, Pardanthus* and *Tigridia*. All are readily propagated by seeds, but this mode is rarely practised except for the purpose of producing new varieties. The seeds should be sown in boxes or small pots, using light, friable soil; the plants are then placed in frames or kept in a greenhouse until the plants are large enough to be safely set out in the open ground. Tender species must, of course, be protected in winter, either by lifting the plants and storing them in a cellar, in pits, or in some similar place where they will not be frozen. The different species of Crocus usually multiply very rapidly by the natural increase of the bulbs or corms. The Gladioluses multiply less rapidly, as a new bulb is only formed at the base of each stem, or set of leaves arising from the buds or eyes on the old or parent bulb, and while there may be many of these buds on its surface, the larger proportion remain dormant, two or three of the strongest pushing into growth. In addition to the buds on the upper surface of a Gladiolus bulb, there are usually produced a large number of bulbils attached to the base of the old bulb, as shown in figure 109. These small bulbs or "spawn," as they are sometimes called, are extensively employed in propagating choice varieties. They should be carefully removed from the old bulbs when the latter are taken up in the fall, and stored in pure sand, and kept where they will not become so dry as to shrivel, nor so moist as to cause them to decay. In spring they should be sown in shallow trenches, and in a very light but rich soil. With good care they will produce bulbs from a half inch to an inch in diameter the first season. But there are some of the cultivated varieties which do not produce the small bulbils in any considerable number, and in some seasons none at all; consequently, the propagator must devise some other mode for rapidly increasing his stock. There are several ways of forcing the latent buds of the old bulbs, and each new sprout will eventually produce a new bulb at its base.

Fig. 109.—GLADIOLUS BULB WITH BULBILS ATTACHED.

HERBS, TUBERS AND BULBS.   325

The old bulbs may be partly divided with the point of a knife, cutting around and at some distance from each eye, found on the top and sides of the bulbs; this will allow the bulb to expand or spread apart as growth begins in spring, each bud producing a shoot or plant. This cutting of the upper surface of the bulb should be done several weeks before the bulbs are planted out, either in frames or the open ground. Separating the bulbs into as many pieces as there are eyes is another

Fig. 110.—GLADIOLUS BULB WITH ROOTS CHANGED TO LEAVES.

mode, but there is danger of the smaller pieces rotting if planted out in cold soil, and the better way is to plant them in shallow boxes of light soil or sand, and keep them in a greenhouse until late in spring, and then transfer to the garden. The roots of the Gladiolus may also be made to change into sprouts, as I had occasion to show in the "American Agriculturist" for June, 1869. A number of Gladiolus bulbs having been left under the stage of a greenhouse during the winter, several of these were turned upside down, so that the eyes or buds were excluded from the light, and probably had less heat than the bottom of the bulb. The results were that the roots, or the appendages which would have been roots under natural conditions, pushed upward in the form of leaves, as shown in figure 110. This is but another instance of the reorganization of cellular matter referred to in a previous chapter. By dividing large, mature bulbs crossways, and inverting the lower section, and planting it in sand, with root surface exposed to the light, sprouts may be obtained in large numbers, each producing a small bulb at its base. The upper half of the bulb may be lightly scarified on the surface, and forced in the same way as though it had not been divided. *Irises*, *Ixias* and *Pardanthuses* are readily propagated by offsets, which are produced very freely. The *Tigridias* have rather small, compact bulbs, many in a cluster, or clump, as shown in figure 111. They must be kept in a warm, dry place during the winter months, and when taken up in the fall the leaves and stems should be left entire, and not cut off until towards spring, or after the bulbs have become thoroughly dried.

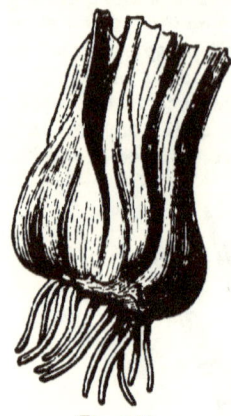

Fig. 111.
BULB OF TIGRIDIA.

Labiatæ (Mint Family).—A very large order mostly of herbs, usually with scented foliage. Many of the genera furnish medicinal and economical plants, such as Sweet Basil, Horehound, Lavender, Mint, Patchouly, Rosemary, Sage and Thyme. These are simple garden herbs, all readily propagated by seed, divisions of the roots, or cuttings of the young shoots planted under glass. The ornamental species of *Salvia* are mostly natives of warm climates, requiring the temperature of a greenhouse in winter; but they are readily propagated by seed, and the varieties are usually perpetuated by young plants struck from cuttings taken from old stock plants kept over for this purpose. The many varieties of *Coleus Blumei*, of Java, now so common in cultivation, are also propagated by cuttings of the young shoots, and even a single leaf may be utilized for this purpose, roots being freely produced if the cuttings are kept moist, and in a temperature ranging from sixty to eighty degrees. There is a tuberous-rooted species of the Coleus (*C. tuberosus*) in Madagascar, and although it is said to have been cultivated for centuries by the natives, as an article of food, it does not appear to have reached this country. The tubers are preserved in a dry, warm place during winter and planted out in spring.

Leguminosæ (Pulse Family).—An extensively and widely distributed order of trees, shrubs, annual and perennial herbs. As the most important of the trees and shrubs of this family have been referred to in preceding chapters, I will notice here only a few genera containing familiar perennial herbaceous plants. Among these are *Astragalus*,

Dalea, Desmodium, Dolichos, Clianthus, Clitoria, Lespedeza, Mimosa, and *Phaseolus.* Of each of these genera there are a few species in general cultivation, either as conservatory or hardy border plants, while some of the genera, such as *Astragalus* and *Desmodium*, are mostly pestiferous or uninteresting weeds. *Dolichos lignosus* is a showy greenhouse climber from India, readily propagated by seed, or cuttings of its perennial roots. *Clianthus Dampieri*, or Glory Pea, is a remarkable showy plant from Australia, and thrives only in a high temperature. Seeds should be sown singly in small pots, and the plants carefully shifted into larger ones as they increase in growth, great care being required in the operation to prevent disturbing, or allowing the soil to fall away from the roots. Plants set out in the garden late in spring will usually bloom the same season. The *Clitorias* are tropical climbing plants, with large and showy flowers, and with trifoliate and pinnate leaves. Propagated by seed, or from cuttings of the side shoots taken off with a hip, and planted in close frames in the house. *Lespedeza* has given us an excellent forage plant for the South in the Japan clover (*L. striata*), and a large late, flowering hardy herbaceous plant, the *L. bicolor*, but usually offered by florists under the name of *Desmodium penduliflorum*. The Japan clover spreads rapidly by seed, and the latter is propagated by dividing the rather hard woody stems and roots. *Mimosa* is the well-known sensitive plant, and is readily propagated by seed. *Phaseolus* contains the annual and perennial beans; the latter may be increased by either seeds or cuttings.

Liliaceæ (Lily Family).—An immense order of about 180 genera and fully 2,500 species. There are also in cultivation innumerable varieties of nearly all of the popular species in the different genera. The space at my command will only admit of a brief notice of a very few of the most familiar genera cultivated for ornamental purposes. Although in intrinsic value such economic plants as the Asparagus, Onion, Squill, and New Zealand Flax (*Phormium tenax*), should take precedence of the purely ornamental, but as luxuries are usually more highly prized than the necessaries of life among civilized nations, so the ornamental must take precedence here of the purely useful among the lilies. Of the latter the *Agapanthus, Fritillaria, Hyacinthus, Lilium,* and *Tulipa*, are the most extensively cultivated and highly prized. The species of all these genera are readily propagated by seed, division, offsets, bulblets—which some species produce in the axils of the leaves—and from scales of the old or mature bulbs. With all the different genera and species having scaly bulbs, such as the *L. Speciosum*, figure 4, Chapter XI., and *L. Canadense*, figure 112, may be readily utilized in their propagation. Imported bulbs, or those which have been a long time out of the ground, or until they have become much shrivelled, may always be used with advantage in this mode of propagation. If such bulbs are planted entire they are very likely to decay, but if the scales are separated and scattered between layers of damp moss, in large pots or well drained boxes, and then placed in a green-house or warm cellar, and given

328   PROPAGATION OF PLANTS.

water as often as necessary to prevent drying, they will usually produce plump little bulbs, in two or three months. When new roots push out from the base of these young bulbs they may be potted separately or pricked out in shallow boxes filled with light rich soil. Hyacinths are rarely propagated in this country; nearly all the bulbs cultivated here are imported from Holland, as they soon degenerate in our climate. The bulbs being solid and not made up of scales, as in the true lilies, the propagation of varieties is effected not only by natural division but by cutting off the upper half of the bulb. This forces the base or lower half to produce a large number of buds or bulblets. Sometimes the lower

Fig. 112.—BULB OF LILIUM CANADENSE.

part is cut across the bottom in various directions and then planted, the exposed parts producing small bulbs near the roots. The young leaves are sometimes utilized in propagation, for if cut off and planted in light soil they will produce bulbs on the lower end. Tulips divide naturally, and increase very rapidly without artificial aid.

Loranthaceæ (Mistletoe Family).—An interesting order of parasitic plants living upon trees and shrubs, and mainly drawing their sustenance from the plants which they infest. There are about fifteen genera and three hundred species known, but the most familiar are the European Mistletoe (*Viscum album*) and the common American Mistletoe *Phoradendron flavescens*. Their propagation is rarely attempted, but if the seeds are inserted just under the bark of the trees the different genera and species are known to live upon, they will usually sprout and grow. The European Mistletoe does not thrive in this country; at least I do not know of any specimens; but the American species grow abundantly from Ohio, south and westward, on the Button-wood, Poplar, Ash, Honey Locust, etc.

Malvaceæ (Mallow Family).—An immense and important order of herbs, shrubs, and trees. There are about fifty genera and nearly, or quite, seven hundred species. All are innocuous mucilaginous plants, with fibrous bark. *Gossypium*, in economic value, is probably the most important genus, yielding the different species and varieties of cotton. The ligneous species of the best known genera have been referred to in preceding chapters, and I will only mention here a few of the hardy herbaceous kinds, such as *Callirrhoë* (Prairie Mallow), *Althea* or *Malva alcea* (Hollyhock), and the *Hibiscus*, or Rose Mallow, of various species indigenous to the United States. The Prairie Mallow (*Callirrhoë*) is readily propagated by seed, and dividing the old plants. The Hollyhock, although naturally a biennial, becomes a perennial in cultivation, through propagating it by division, and by cuttings of the young sprouts as they push from the large fleshy roots in spring. The Rose Mallows are not often cultivated, but are worthy of more attention than they have received. New varieties are readily raised from seed sown in a half-shady bed, in the open ground in spring, and these propagated by division or young cuttings.

Melastomaceæ (Melastoma Family).—An order of trees, shrubs and herbs. They are most abundant in the tropics, and representatives of several genera are grown in greenhouses. Among these are: *Centradenia, Cyanophyllum, Bartoloma, Melastoma* and *Pleroma*. The most familiar native genus is the *Rhexia*, half a dozen species of which are found in the Atlantic States, but rarely seen in cultivation. Propagated by seed and cuttings in summer, planted in a close frame or under a hand-glass.

Mesembryanthemeæ (Fig Marigold Family).—This family is also known as the *Ficoideæ* in some of our botanical works, and in others as *Mesembryaceæ*. There are nearly three hundred species in cultivation, mostly of the one genus *Mesembryanthemum*. They are best known in this country under the popular name of "Ice Plants," They are chiefly low trailing plants, with thick fleshy leaves, and some of the species have large and showy flowers. They thrive best in a rather light poor soil, with more or less lime rubbish intermixed with it. Their propagation is exceedingly simple, as almost any piece of stem or leaf will take root if laid on the surface of sand or thrust into it, and then exposed to the direct rays of the sun.

Nymphæaceæ (Water Lily Family).—A rather small order of aquatic perennial herbs, widely distributed over the globe in freshwater ponds, and along the borders of rivers and smaller streams. The leaves and flowers of some of the genera, like those of the common Water Lily (*Nymphœa odorata*), float on the surface; others, such as *Nelumbium luteum* and *speciosum*, rise several feet above the surface, the seed vessels ripening in this position. In others the seeds or fruit mature under water. They thrive best in rather shallow ponds or streams with muddy bottoms, but deep enough to prevent the roots freezing. Propagated by seed and division of the roots, or by tubers formed on the subterranean stems and root stocks. The seeds may be sprouted in water in a greenhouse, and then transferred to the pond, a small stone or pebble being attached to each to sink and hold it in position at the bottom. I have always found in planting the tubers and cuttings that it was better to sink each with a stone than to trust to their own weight to carry them down.

Onagrarieæ, or *Onagraceæ* (Evening Primrose Family).—A large order, mostly inodorous, of annual and perennial herbs; rarely shrubs or trees. The most familiar genera in cultivation are: *Clarkia* (annual), *Fuchsias* (shrubs), *Jussiæa*, *Gaura*, *Œnothera* and *Zauschnera*. The shrubby species are propagated by seeds and cuttings, and the herbaceous perennials by seeds and careful divisions; or by cuttings of the young shoots, taken off early in spring and planted in a close frame where they will receive but a moderate heat. The *Œnotheras* are very showy plants, thriving best in light dry soils.

Orchidaceæ (Orchis Family).—An immense order of monocotyledonous perennial herbs. Some are terrestrial, with tuberous or fascicled roots; others are epiphytes, with or without pseudo-bulbs, living on the stems of palms, and various kinds of trees and shrubs in tropical countries. The flowers are of peculiar and varied structure; some of them appear more like birds and butterflies than flowers. Of the terrestrial genera there are a large number indigenous to the United States, but they are rarely cultivated, with the exception of the different species of *Cypripedium*, or Lady's Slipper. These thrive best in moist soil, and the plants may be obtained from their native habitats, transplanting when the flowers begin to fade, or late in summer. The tender exotic species are cultivated in greenhouses, in pots filled with light fibrous or peaty soil, water being plentifully supplied when the plants are growing most rapidly, and the temperature varied according to the requirements of the different species. There are three hundred and thirty-four genera in the Orchis Family, and about five thousand species. The genera are divided into five tribes—*Epidendreæ*, *Vandeæ*, *Neottieæ*, *Ophrydeæ* and *Cypripedieæ*. The genera of the first three tribes are mostly epiphytal, and in cultivation are grown in wire baskets filled with sphagnum, bits of cork, light wood, bark, and similar materials; or they are attached to sheets of cork, sections of the stems of old palms, lumps of charcoal; in fact, almost any porous substance which will not decay rapidly will answer. Ordinary flower pots will answer for many of the species if filled with soft brick and pieces of cork or porous wood. A pot designed expressly for Orchids has recently been

HERBS, TUBERS AND BULBS. 331

introduced by Mr. Matthews of England, and its construction is clearly shown in figure 113. A circular earthenware disc is made to replace the mass of broken crocks usually placed below the other materials for drainage, thereby securing aëration and avoiding all danger of overwatering. Tribes four and five are terrestrial Orchids, and grown in light soils as recommended for the common Lady's Slipper. Few of the Orchids are of any economic value; the Vanillas, however, are an exception, the fruit yielding a valuable balsamic oil with well known delicious perfume. The propagation of the Orchids, as a family, is a rather slow process, and while species of the epiphytal tribes may be successfully divided when at rest, yet cultivators depend mainly upon fresh

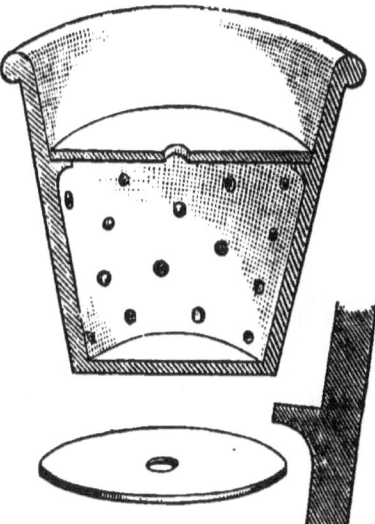

Fig. 113.—ORCHID POT.

collections from the native habitats of the different species for increasing their stock of plants. New varieties and hybrids are raised from seed, but it requires skill, patience, and structures especially adapted to the needs of these plants, to enable one to succeed in such operations.

Papaveraceæ (Poppy Family).—A small order of about seventeen genera, mostly herbs with milky or colored juice. The only shrub belonging to this family is *Dendromecon rigidum* of California. There are several genera containing some showy annuals that have long been familiar in gardens, like *Eschscholtzia*, *Argemone*, and the Opium Poppy (*P. somniferum*). But the perennial Poppies (*P. orientalis*) are not so well known, but they are well worthy of cultivation. They are easily raised from seed, but rather difficult to transplant, owing to their very long and deeply penetrating roots. The seeds should be sown as soon as ripe, and where the plants may be left undisturbed.

Plumbaginaceæ (Leadwort Family).—These are mostly low-growing marine herbs, somewhat remarkable for their regular flowers; all the different parts or organs in fives—five lobes to the calyx, five petals to flowers, and the same number of stamens and pistils. The most familiar genera in gardens are *Armeria* (Sea Pink and Thrift), *Statice* (Sea Lavender) and *Plumbago* (Leadwort). They are plants requiring only moderate care and are easily propagated by seed, or by dividing the clumps or roots.

Portulacaceæ (Purslane Family).—Succulent annual and perennial herbs. The common garden Pusley or Purslane (*P. oleracea*), and the Great-flowered Portulaca (*P. grandiflora*), are well-known annual representatives of this family. There are very few species of the other fourteen genera in cultivation, except the two species of Claytonia (*C. Virginica* and *C. Carolina*), and these are not very common in gardens, although well deserving of a shady nook, even in grounds of limited extent. All readily propagated by seeds.

Primulaceæ (Primrose Family).—An order of about twenty genera and 150 species of low-growing herbs of no economic value, but a few of the genera, such as *Androsace*, *Cyclamen*, *Dodecatheon* and *Primula*, are highly-prized ornamental plants. The *Androsaces* are all Alpine plants, and thrive only in open, airy situation, among rocks and stones, and in positions where they will not suffer for moisture. Of the *Cyclamens*, there are numerous species, but the varieties in general cultivation as greenhouse plants are offsprings of *C. Persicum*. These have thick fleshy root-stocks or corms, which are largely imported for forcing under glass. They are usually propagated by seeds sown as soon as ripe, in shallow boxes or seed-pans, and in frames or in an ordinary greenhouse. *Dodecatheon*, or American Cowslip, is a genus of only one species, but there are a large number of natural local varieties, widely distributed from our Middle States to California, and northward to Washington Territory. They thrive best in moist or wet soils, and half shady positions. They are handsome plants for forcing in a cool greenhouse, and readily propagated by seeds, or by dividing the crowns. The genus *Primula* contains many species, chiefly native of the cold regions of Europe and Asia, but the most popular varieties for greenhouse culture belong to a Chinese species (*P. Sinensis*). They are plants requiring simple culture, and are propagated by seeds, cuttings of the roots, and by dividing the old plants as soon as they have ceased blooming.

Ranunculaceæ (Crowfoot Family).—An immense order, and one largely represented in cultivated plants. Mostly herbaceous or somewhat shrubby, with acrid, caustic, and in some genera, very poisonous juice. The *Anemone* (Wind-Flower), *Actæa* (Baneberry), *Aconitum* (Monkshood), *Aquilegia* (Columbine), *Clematis* (Virgin's Bower), *Delphinium* (Larkspur), *Helleborus* (Christmas Rose), *Hepatica* (Liverleaf), *Pæonia* (Pæony) and *Ranunculus* (Buttercup), are some of the best known genera. The indigenous, and many of the exotic species are hardy garden plants, of easy culture, and readily propagated by seeds and division of the roots. The double garden *Ranunculuses*, descended from *R. Asiaticus* and *R. aconitifolius*, are but sparingly cultivated in this country, as they are not hardy, and if planted early in spring the foliage and flowers are burned by the sun when they come into bloom.

Resedaceæ (Mignonette Family).—A small order of herbs, or slightly shrubby plants, rarely cultivated except species of Mignonette (*Reseda*). These, in the *R. odorata*, a native of North Africa, have small, inconspicuous, but sweet-scented flowers. There are many varieties in cultivation, and propagated by seeds, and by cuttings of the young shoots.

Rutaceæ (Rue Family).—Mostly trees and shrubs, characterized by their strong smell, as in the Orange and Lemon, and our common Prickly Ash. There are a few genera of hardy herbaceous perennials, such as the *Dictamnus Fraxinella* and *Ruta graveolens*, or garden Rue. Readily propagated by seeds, or by dividing the roots late in autumn or early spring.

Sarraceniaceæ (Pitcher-Plant Family).—A small order of bog plants, with pitcher-shaped, tubular and hooded leaves, with rather large, six-petaled flowers. There are half a dozen species of the *Sarracenias* native of the Atlantic States, and only one of the genus *Darlingtonia*, and this a native of California. Propagated by seed sown in very light soil or on sphagnum, and by dividing the roots.

Saxifragaceæ (Saxifrage Family).—A very large family, of about seventy-five genera, and five or six hundred species. They are mostly shrubs, and the Currant, Gooseberry, *Deutzia* and *Hydrangea* are familiar representatives. Among the perennial herbaceous genera, the *Astilbe* and *Saxifraga* are the most common and best known. The *Astilbe Japonica* (often incorrectly called *Spiræa Japonica*) is a very hardy border plant, extensively cultivated by florists for forcing in winter, its small, pure white flowers being in great demand. The Beefsteak Saxifrage (*S. sarmentosa*) is an old and well-known trailing greenhouse plant, sometimes called Strawberry Geranium. The Siberian Saxifrages (*S. crassifolia*), and its many varieties, are occasionally seen in gardens. All readily propagated by dividing the roots.

Scitamineæ (Banana Family).—An order of tropical and sub-tropical plants, mostly having very large leaves with a distinct petiole and blade. The most familiar representative of this family in gardens is the common Indian Shot plant, or *Canna Indica*, and other species of this genus. *Maranta zebrina* is a well-known and common greenhouse plant. The Queen plant, or *Strelitza reginæ*, from the Cape of Good Hope, is a large, rather coarse, stemless greenhouse plant, also common; while the Banana (*Musa*), of which there are several species, are occasionally cultivated in conservatories, but it is a large, coarse-growing plant, requiring considerable space for full development. They are propagated by dividing the roots. The Cannas are most extensively cultivated, as they are excellent plants for bedding out in summer. The thick fleshy roots should be stored in a warm, dry place in winter. New varieties are raised from seed, which should be soaked in warm or tepid water for a day or two before sowing.

Scrophulariaceæ (Figwort Family).—A very large and widely distributed order of herbs, shrubs, and occasionally trees. The flowers are usually two-lipped, and with an irregular monopetalous corolla. *Antirrhinum* (Snapdragon), *Calceolaria* (Lady's Slipper), *Chelone* (Turtlehead), *Digitalis* (Foxglove), *Maurandia*, including *Lophospermum*, *Mimulus* (Monkey Flower), and *Pentstemon*, are well-known representatives of this family. Varieties are easily raised from seeds, and are then propagated by cuttings of the succulent shoots planted under glass; or, with the hardy genera, by dividing the roots when the plants are at rest.

Solanaceæ (Nightshade Family).—A very large order, mostly herbs, with rank-scented foliage, often containing narcotic or poisonous properties, as in *Nicotiana* (Tobacco), *Atropa* (Belladonna), and *Datura* (Stramonium). Several genera have given us very valuable economic plants, such as the Tomato, Egg Plant, Pepper (*Capsicum*), Potato, and Tobacco. Among those genera cultivated principally for ornamental purposes, the *Petunias*, *Nierembergias* and *Cestrums*, are the most common in gardens and greenhouses. They are readily propagated by green cuttings under glass, and the tuberous rooted by divisions. The best and most simple method of perpetuating varieties of the Tomato is by cuttings, which strike root very readily in frames during the summer or in the greenhouse in winter.

Umbelliferæ (Parsley Family).—A large family, mostly innoxious, biennial or perennial herbs, but a few are poisonous, such as the Poison Hemlock (*Conium maculatum*), the Water Hemlock (*Cicuta maculata*), and the Water Parsnip (*Sium lineare*). But this family contains many genera of great economic value, such as the common garden Carrot, Caraway, Coriander, Fennel, and Parsnip. The cultivation and propagation of these plants are too well known to be considered in a work of this kind.

Urticaceæ (Nettle Family).—A large order of several sub-families, such as the Elm family, Fig family, etc. The true Nettles are annual or perennial herbs with a fibrous bark, as in the *Bæhmeria nivea*, or Ramie plant, of China, which yields one of the most valuable of textile fibers, and is woven into grass cloth. It is not hardy in our Northern States, but is now being extensively cultivated in our Southern. It is propagated by seed, cuttings of the young shoots and of the subterranean stems. The Hop plant also belongs to this family, and is usually propagated by dividing the roots, or by removing the young sprouts with a few roots attached.

Valerinaceæ (Valerian Family).—A small order of herbs, and only a few genera represented among cultivated plants. The common garden Valerian (*Valeriana officinalis*), is usually cultivated as a medicinal plant; its flowers are very sweet scented, and its roots have a very strong odor.

An indigenous species (*V. edulis*), has large spindle-shaped roots, and is said to have been used as food by the Indians. *Centranthus ruber*, or Spurred valerian, has red flowers, and sometimes cultivated in European gardens under the name of Jupiter's Beard. Readily increased by seed or by dividing the roots.

Verbenaceæ (Vervain Family). — A small order of herbs and shrubs; the latter have already been noticed in previous chapters. See *Callicarpa*, *Lantana*, and *Lippia*. Of the true *Verbenas* there are a large number of species, but those most usually cultivated in gardens are descended from several South American species, but now so mixed that it would be impossible to trace them back to the original types. New varieties are raised from seed, and then propagated by layers or cuttings of the young shoots. The latter mode is the preferable one, and for stock plants to be preserved over winter, the cuttings should be struck in August or September, and then kept in a rather cool house until toward spring, when they are forced for the purpose of producing more new shoots for cuttings.

Violaceæ (Violet Family).—This family is scarcely represented in cultivation by any other genus than *Viola*, the true Violets and Pansies. The most popular species is *V. odorata*, of which there are many varieties in cultivation. Some are quite hardy, others tender and need slight protection and are usually grown in cold frames where the temperature may be under control of the gardener. There are about thirty North American species of *Viola*, but they are mostly scentless, or only slightly fragrant. The Pansy or Heart's-ease, is descended from *V. tricolor* of Europe, and has been greatly improved, and the flowers increased in size by cultivation. The Violets thrive best in a very rich, light soil, and with a moderate temperature. Plants of easy culture and all readily propagated by seed, cuttings, or dividing of the roots.

INDEX.

Abelia 249
Abies 233, 237, 249
    Pectinata 237
Abronia 249
Abutilon 233, 249
    Experiments with 123
    Fertilization of 105
    Species of 123
Abutilon, Boule de Neige and
    Santana 123, 125
    Megapotamicum 233
    Santana 123
    Thomsonii Pleno 111
Acacia 250
Acalypha 250
Acanthaceæ 311
Acanthus 311
Acer 233, 250
    Campestre 234
    Dasycarpum 233, 250
    Macrophyllum 76
    Negundo 250
    Pseudo-Platanus 233
    Rubrum 233, 250
Achania Malvaviscus 282
Achillea 317
Achimenes 323
Aconitum 332
Acorns 73
Actæa 332
Actinidia 250
Adder's Tongue Fern 320
Adenocalymna 251
Adenocarpus 251
Adenostoma 251
Adhatoda 251
Advantages and Disadvantages
    of Layers 175
Aerial Organs of Stems 60
Æchmea 314
Ægiceras 251
Æsculus 251
    Californica 234, 252
    Glabra 234
    Hippocastanum 234
    Parviflora 252
Aganosoma 252
Agapanthus 327
Agapetes 252

Agathophyllum 252
Agathosoma 252
Agave 311
Ailanthus 252
Akebia 252
Alder 234
Alhagi 252
Almond 224, 253
Alnus 234, 253
    Cordifolia 234
    Firma 252
    Glutinosa 234, 252
    Oblongifolia 234
    Rubra 234
Alocasia 312
Aloysia Citriodora 281
Alternanthera 311
Althea 276, 329
Alyssum 318
Amaranthaceæ 311
Amaranth Family 311
Amaryllidaceæ 311
Amaryllis Family 311
Amelanchier 230, 252
American Aloe 311
    Azaleas 235
    Burning Bush 230
    Cowslip 232
    Holly 241, 277
    Linden 244
    Persimmon 231
    Sweet Birch 235
    Sweet Chestnut 236
    Weeping Willow 243
    White Fringe 237
    Yew 300
Amorpha 253
Amorphophallus 312
Amsonia 312
Amygdalus 253
    Nana 253
Ananassa 314
Anchusa 313
Andromeda 253
Androsace 332
Angers Quince 232
Anemone 332
Anise Tree 278
Anona 253

INDEX. 337

Anona Cherimolia............253
    Muricata................253
    Squamosa...............253
Antennarias..................317
Anthemis nobilis.............217
Anthurium....................312
Antirrhinum 334
Apocynaceæ..................311
Apocynum 312
Apple........................224
    Gall 27
    Seedlings 37
Apricot......................224
Aquilegia 332
Arabis.......................318
Araceæ 312
Aralia.......................253
    Chinensis 253
    Maudchuricus...........253
Araucaria....................254
Arbor-Vitæ 234, 301
Arbor-Vitæs 238
Arbutus 234, 254
    Unedo 234
Arctostaphylos...............254
Ardisia......................254
Argemone.....................331
Armeria......................332
Aristolochia.................254
    Sipho 254
Aroidæ 312
Artemsia 317
Arrow Wood..................302
Arthrotaxis 235
Artichoke....................316
Artocarpus...................254
    Incisa..................254
Arum Family 312
Ash 240, 274
    American 240
    European 240
    Seeds 75
Ash-leaved Maple............127
Asimina......................254
    Triloba............227, 224
Asparagus 327
Aspen 280
Aster 317
Astilbe Japonica.............333
Astragalus...................326
Atragene.....................255
Aucuba.......................255
Auraucaria...................234
Australian Gum Tree..........271
Austrian Pine................237
Azalea 234, 255
    Calendulacea...........235
    Indica 255
    Mollis 235, 255

Azalea Nudiflora.............234
    Pontica................235
    Sinensis 235, 255
Azara........................255
Baccharis....................255
    Halimifolia 256
Bald Cypress.................200
Balsam Fir...................249
Bamboo.......................323
Bambusa......................323
Banana.......................333
Banana Family...............333
Baneberry....................332
Banking up Sprouts...........273
Banksia 256
Barberry 256
Bartoloma 329
Bastard Oak Apple............ 26
Bass 189
Basswood................244, 301
Bay Tree.....................280
Bear Berry...................254
Beech 212, 240
    American 240
    European 240
Beefsteak Saxifrage..........333
Begoniaceæ...................313
Begonia Family...............313
    Leaf Cutting of.........167
Bellflower Family............315
Bell-glass 160
Benthamia 256
    Japonica 256
Bent Layer of Vine...........173
Berberis 256
    Canadensis..............256
    Vulgaris 256
Betula 235, 256
    Alba 235
    Lenta 235
    Papyracea...............235
Bignonia 257
    Capreolata 275
    Radicans................257
Bilbergia....................314
Bindweed 318
Biotas 238, 301
Birch 235, 256
Bird Cherry..................226
Birthworts...................254
Bitter-sweet 263
Blackberry 296
Black Damas Plum.............231
Bladder-nut..................299
Bladder Pod..................279
Bladder Senna................266
Bleeding Heart...............321
Blueberry 302
Bœhmeria Nivea...............334

338 PROPAGATION OF PLANTS.

Boraginaceæ 313
Borbonia 257
Bordguer's Grafting Tool 200
Borage Family 313
Boronia 257
Botrychiums 320
Bottle Grafting 219
Bouvardia 257
Bovista Gigantea 15
Box Tree 258
Bow-Wood 281
Bread Fruit 254
Bridal Wreath 298
Bromelia 314
Bromeliaceæ 314
Broom 239, 269
   Irish 239, 269
   Scotch 239, 269
   Spanish 239, 269
Broussonetia 258
Brugmansia Candida 270
Bryonias 316
Bryophyllum 318
Bucco 252
Buckeye 234
Buckthorn 292
Bud, Incision Ready for 161
   In Position 191
   Variation 132
Budding with Sloping Cut 194
Buds, Different Forms of 29
Buffalo Berry 298
Bulb of Lilium Canadense 327
   of Lilium Speciosum 23
   of Tigridia 326
Burning Bush 239
Buttercup 332
Butternut 241
Buttonwood 288
Buxus 258
Cabbage 318
Cactaceæ 314
Cactus Family 314
Caladium 312
Calceolaria 334
Calico Bush 279
California Buckeye 252
California Nutmeg 301
Calla 312
Calla Lily 312
Callicarpa 250, 335
Callirrhöe 324
Callistemon 258
Callistephus 317
Callitris 258
Calluna 258
Calophaca Wolgarica 235
Calothamnus 259
Calycanthus 185, 259

Calystegia 318
Camellia 235, 259
   Theifera 259
Campanulaceæ 315
Campanula 315
Camperdown Weeping Elm ... 244
Camphor Tree 280
Candytuft 318
Canna Indica 333
Cape Jessamine 274
Caper Family 315
   Tree 259
Capparis 315
   Spinosa 234
Capparidaceæ 315
Carpinus Betulus 236
Caprifolium 281
Capsicum 334
Caragana 235
   Arborescens 259
   Arborea 241
Caraway 334
Carbon 46
Carex 319
Carnations .... 315'
Carob Tree 264
Carrot 334
Carpinus Americana 236
Carya 259
Caryophyllaceæ 315
Cassandria 262
Cassia 262
Castanea 286, 262
   Vesca, var Americana ... 236
Catalpa 263
   Bignonioides 236
   Speciosa 236
Cauliflower 318
Ceanothus 263
Cedar 236, 263
   of Lebanon 238
Cedars 238
Cedrus 236, 238, 263
   Atlantica 238
   Deodara 238
   Libani 238
Celastrus 263
Celosia 311
Celtis 263
Centradenia 329
Centranthus Ruber 335
Cerastium 315
Cerasus 236, 289
   Avium 225
   Capronianum 225
   Demissa 226
   Ilicifolia 237
   Lauro-Cerasus 237
   Lusitanica 237

Cerasus Padus..............226	Claytonia Virginica..........332
Pennsylvanicum 226	Clematis 332
Pumila 226	Cleome 315
Serotina 226	Clerodendron 265
Virginiana 226	Clethra 265
Ceratiola....................264	Cleyera 266
Ceratonia 264	Clianthus 327
Cercidiphyllum 264	Clitoria 327
Cercis 264, 314	Cockscomb Coral Tree 271
Cestrums 334	Coco Plum..................265
Chamæcyparis 237, 264	Codiæum 266
Lawsoniana...............264	Coleus Blumei 326
Nootkaensis..............264	Tuberosus 326
Thyoides..................264	Colocasia 312
Chamiso 251	Esculenta 312
Chamomile..................317	Columbine 332
Chaste Tree..................302	Colutea 266
Chelone.....................334	Comfrey 313
Cherry 224, 236, 263	Commelina.................316
Cherimoyer 253	Commelinaceæ 316
Cheiranthus 318	Common Budding Knife.......188
Chestnut 236, 262	Common Plum 70
Chilopsis....................265	Compositæ..................316
China Aster.................217	Composite Family............316
China Tree..................282	Comptonia 266
Chinese and Japan Cherries....226	Cone-Bearing Trees 237
Aralia 253	Coniferæ....................237
Azaleas 235, 255	Conifers 237
Fir 268	Conium Maculatum 334
Pæony 285	Convolvulaceæ 318
Quince 232, 269	Convolvulus Family.........318
Snowball..................302	Coontie 319
Yam 319	Coral Tree 271
Chionanthus 265	Corchorus 279
Retusus 237	Cordia 266
Virginica..................267	Corema 266
Choisya 265	Coriander 334
Choke Cherry 226	Cornel 266
Christ's Thorn...............285	Cornel Tree 226
Chrysanthemum..............317	Cornus.....................266
Chrysobalanus 265	Florida............226, 239, 266
Cinchona 286	Florida Pendula...........266
Cicuta Maculata 334	Mas......................266
Cinnamon 280	Florida Purpurea..........267
Vine 319	Corydalis...................322
Cinquefoil 289	Corylus 267
Cion and Stock...............205	Avellana 239
Cion of American Holly 277	Cotoneaster.............239, 267
Circulation of Sap............85	Cottonwoods................259
Cistaceæ 316	Crab Cuctuses..............314
Cistus 276, 316	Cranberry 302
Citron 226	Cranberry Tree..............302
Citrus 265	Crape Myrtle................280
Decumana 226	Crassula 318
Cladrastis 265	Crassulaceæ 318
Tinctoria................265	Cratægus 239, 267
Amurensis 265	Coccinea 230
Clarkia......................330	Cratæva 315
Claytonia Carolina............332	Cress 318

Crinum 311
Crocus 324
Cross Fertilization 107
Croton 266
Crowfoot Family 332
Cruciferæ 318
Cryptomeria 268
Cucubitaceæ 319
Cucumber Tree 282
Cunninghamia 268
Cupressus 268
Currant 226, 293
Cuscuta 318
Custard Apple 254
Cut-leaved Chaste Tree 302
Cutting Below a Bud 149
   of Green Wood 102
   of Hollow Stem 149
Cuttings of Immature Growths .. 154
   of Leaves 165
   of Mature Growths 144
Cyanophyllum 320
Cycadaceæ 319
Cycas Family 319
   Revoluta 319
Cyclamen 332
Cydonia 232, 268
   Chinensis 269
   Japonica 268
   Vulgaris 268
Cynips Inanis 26
   Spongifica 25
Cyperaceæ 319
Cyperus 319
Cypress 237, 264, 268
   Vine 318
Cypresses 238
Cypripedieæ 330
Cyrilla 269
Cytisus 239, 269
Dacrydium 269
Dalea 327
Daphne 269
   Cneorum 269
   Laureola 239
   Mezereum 269
Darlingtonia 333
Darwinia 269
Dasylirion 269
Date Plum 227
Datura 270, 331
   Arborea 270
   Meteloides 270
Deciduous Cypress 300
Decumaria 270
Delphinium 332
Dendrocalamus 323
Dendromecon Rigidum 331
Desert Willow 265

Desfontanea 270
Desmodium 327
   Penduliflorum 327
   Striata 327
Deodar Cedar 238
Deutzia 333
   Gracilis 270
Dianthus 315
Dicentra 321
   Canadensis 321
   Cucullaria 321
   Spectabilis 321
Dicotyledonous Plants 55
Dictamnus Fraxinella 333
Diervilla 270
   Sessiliflora 270
   Trifida 290
Diffenbachia 312
Digitalis 334
Dimorphanthus 253, 270
Dicecious Plants 68
Dionæa 320
Dioscoreaceæ 310
Dioscorea Batata 319
   Sativa 319
Diospyrus Kaki 231
Distribution of Seeds 75
Dodder 318
Dodecatheon 332
Dogbane Family 311
Dog Rose 243
Dogwood 238, 266
Dolichos 327
Double Abutilon 111
   Bell-Glass 169
   Flowers 110
   Pot for Cuttings 168
   Worked Trees 230
Doucin Apple 224
Douglass Spruce 290
Downing Mulberry 283
Drosera 320
Droseraceæ 319
Drosophyllum 320
Duplicated Corolla 111
Dutchman's Breeches 321
   Pipe 254
Dwarf Cherry 226
   Double Flowering Almond .. 253
East India Mango 282
Echinocactus 314
Egg-Plant 334
Elæagnus 270
Elder 297
Elecampane 317
Elephant's Ear 313
Elliottia 270
Elm 244, 302
   Seeds 74

INDEX. 341

Endosmosis ... 19
Endogens and Exogens ... 53
English Bitter Willow ... 243
   Elm ... 244
Epacris ... 270
Epiphyllum ... 314
Epidendreæ ... 330
Eriobotrya Japonica ... 286
Erica ... 270
Erythrina ... 271
   Crista-galli ... 271
   Herbacea ... 271
Eschscholtzia ... 231
Eucalyptus ... 271
   Globulus ... 271
Euphorbia ... 272
Euonymus ... 271
   Atropurpureus ... 239
   Europæus ... 239
   Japonicus ... 271
   Latifolius ... 240, 272
   Radicans ... 271
European Holly ... 241
   Plum ... 231
   Sticky Alder ... 252
Evening Primrose Family ... 330
Everlastings ... 317
Evergreen Beeches ... 240
Exochorda ... 240, 272
Exosmosis ... 17
Experiments with Abutilons ... 123
   with Leaves ... 97
   with Seeds ... 81
Fabiana ... 272
Fagus ... 240, 272
   Ferruginea ... 240
   Sylvatica ... 240
False Acacia ... 293
   Larch ... 238
Faramea ... 272
   Odoratissima ... 272
Fennel ... 334
Fern Family ... 320
   Seedling ... 321
Fertilization of the Kalmia ... 102
   of Orchids ... 122
Fever Tree ... 271
Ficoideæ ... 329
Ficus ... 273
   Carica ... 273
   Elastica ... 273
   Indica ... 273
Fig ... 227
Fig-Marigold Family ... 329
Fig Tree ... 273
Figwort Family ... 334
Filbert ... 239, 267
Field Maple ... 283
Filices ... 320

Firming the Soil ... 153
Fitzroya ... 273
Five Finger ... 289
Flowers, Fruits and Seeds ... 66
   of the Grape ... 108
   of Kalmia ... 102
   Parts of ... 67
Flowering Maples ... 249
Flute Budding ... 196
Fontanesia ... 273
   Fortunei ... 273
   Phillyræoides ... 273
Fontenay Quince ... 232
Food of Plants ... 45
Forsythia ... 273
   Fortunei ... 273
   Suspensa ... 273
Fothergilla ... 273
   Alnifolia ... 273
Foxglove ... 334
Fraxinus ... 240, 274
   Excelsior ... 240
Fremontia ... 274
   Californica ... 274
French Mastic ... 203
   Mulberry ... 268
   Tamarisk ... 300
Fritillaria ... 327
Fruit, Forms of ... 70
Fuchsia ... 230, 274
   Arborescens ... 274
Fumariaceæ ... 321
Fumitory Family ... 34
Gaillardia ... 317
Galanthus ... 311
Gall-Gnat ... 28
Garden Basket ... 318
Gardenia Florida ... 274
Garland Flower ... 269
Garlic Pear ... 315
Garrya ... 274
Gaura ... 330
Gazania ... 317
General Principles of Propagation ... 135
Genista ... 274
Gesneria Family ... 323
Gesneriaceæ ... 323
Gentian Family ... 322
Gentianaceæ ... 322
Genus Hybrids ... 116
Georgia Bark ... 286
Geraniaceæ ... 322
Geranium Family ... 322
Germination of Seeds ... 83
Ghent Azaleas ... 255
Ginkgo ... 207
Gladiolus ... 324
   Bulb ... 324

Gladiolus with Roots Changed to Leaves............325
Gleditschia............240, 274
   Triacanthos............240
   Seed............142
Glory Pea............327
Gloxinia............323
Glyptostrobus............238, 300
Golddust Tree............255
Goat Plant............251
Golden Bell............273
   Chain............279
   Corchorus............279
Gomphrena............311
Gooseberry............227, 298
Gossypium............329
Gourd or Cucumber Family....319
Grafting............205
   Bottle............220
   Modified............220
   By Approach............221
   Cleft............204, 205
     With two Cions............206
   Crown............206, 207
   Saddle............210
   Side-Crown............208
   On Roots............210
   Side, with Vertical Cleft...220
   The Dahlia............317
   The Grape............307
   Triangular Crown............207
   Veneer............214
   Dogwoods............216
   Maples............216
Grafting Wax............201
Gramineæ............323
Grape............302
   Cutting............303
   Cuttings of Green Wood...305
   Layering of............306
   Single-Bud Cutting............304
   Vine Tendril............60
Grass Family............323
Great-Flowered Spiræa............240
Great Tree of California............297
Groundsel Tree............255
Growth of Cells............14
Gymnocladus............275
   Canadensis............142, 275
Hæmanthus............311
Halesia............240, 275
   Tretraptera............240
Halimodendron............241, 275
Hamamelis............275
   Japonica............275
Hand Glass............160
Hand Pruning Shears............186
Hawthorn............239, 267
Hazelnut............239, 267

Heartsease............335
Heart-Shaped Leaved Alder...234
Heath............271
   Cuttings of............164
Heather, Ling............258
Hedgehog Cactus............314
Helianthemum............275, 316
Helleborus............332
Heliotrope............513
Heliotropium............313
Hemlock Spruces............238
Herbs, Tubers and Bulbs............309
Hepatica............332
Hesperis............318
Hercules Club............253
Hibiscus............276, 329
   Rosa Sinensis............276
   Syriacus............276
Hickory............260
   Hale's Paper-Shell............261
   With Root Sprouts............262
Hippeastrum............311
Holly............241, 277
Hollyhock............329
Honey Locust............240, 274
Honeysuckle............281
Hop-plant............334
Hop-hornbeam............284
Hop-tree............290
Horehound............326
Hornbeam............236, 259
Horse-chestnut............251
   Germinating............251
Horseradish............318
Horse Sugar............299
House Leek............318
Hovenia Dulcis............276
Huckleberry............302
Hyacinthus............327
Hybrid Grapes............130
   Raspberries............116
Hydrangea............233, 276
   Paniculata Grandiflora....276
   Quercifolia............276
Hydrogen............45
Iberis............318
Ice Plant............329
Idesia Polycarpa............276
Ilex............241
   Aquifolium............241, 277
   Opaca............241, 277
Illicium............278
Imperfect Bunch of Grapes...109
Imperfectly Fertilized Ear of Corn............106
Inarching............221
Indian Bean............236, 263
   Corn............323
   Currant............299

Indian Hawthorn	292
Hemp	312
Shot Plant	333
Indigofera	278
Indigo Shrub	253
Influence of Cion and Stock	243
of Cion on the Stock	247
of Pollen	117
of Stock on Cion	245
Inula	317
In What Materials to Plant Cuttings	160
Ipomœa	318
Iresine	311
Iridaceæ	324
Iris Family	324
Iron	49
Iron-wood	284
Ixia	324
Japan Alder	252
Arbor-Vitæs	264
Cedar	268
Cherry	276
Chestnut	233
Clover	327
Gooseberry	250
Holly	284
Hydrangea	276
Maples	234
Quince	232, 268
Styrax	240
Jasminum	278
Jessamine	278
Judas Tree	264
Juglans	241, 278
Californica	278
Cinerea	241, 278
Nigra	278
Regia	241, 278
Rupestris	278
Juneberry	230, 252
Juniper	278
Junipers	238
Juniperus	238, 278
Virginiana	238
Jupiter's Beard	335
Jussiœa	330
Kalmia Latifolia	279
Kentucky Coffee Tree	142, 275
Kernel of Walnut	79
Kerria Japonica	279
Kilmarnock Willow	243
Kolreuteria Paniculata	279
Labiatæ	326
Laburnum Vulgare	235, 279
Ladies' Eardrops	274
Slipper	330, 334
Lagerstrœmia	280
Lantana	280, 335

Larch	280
Larches	238
Large-flowered Spiræa	272
Larix	280
Europœa	238
Larkspur	332
Lath Screens for Frames	159
Laurel	263, 279, 280
Laurus	280
Nobilis	280
Sassafras	297
Lavender	326
Lawson Cypress	264
Layered Branch of a Tree	171
Layer in a Pot	174
Layers of Vines	172
Lead Plant	253
Leadwort Family	332
Leaf of Acacia	65
of Buckeye	64
of Begonia	166
of Beech	63
of Cut-leaved Birch	63
of Fern-leaved Aralia	65
of Jersey Pine	62
of Lilac	63, 165
of Locust	65
Leather Leaf	262
Leaves Absorbing Moisture	97
of Larch	63
of White Pine	58
Lefort's Liquid Grafting Wax	203
Leguminoseæ	326
Lemon	226, 265
Lentil Shrub	235
Lespedeza	327
Leyder's Grafting Implement	200
Liatris	317
Libocedrus	301
Licmoophra Splendida	13
Lilac	299
Ligustrum Vulgare	280
Liliaceæ	327
Lilium	327
Canadense	327
Candidum, Flower of	103
Speciosum	327
Lily Bulbs on the Flower Stem	24
Family	327
Tree	268
Lime	49, 226
Limits of Cross-Fertilization	112
of Vitality in Seeds	77
Limonia	280
Trifoliata	226
Linden	244, 301
Lippia	281, 335
Liquidamber	281
Liriodendron Tulipifera	281

Liverleaf	332
Locust, or False Acacia	243
Borer	293
Tree	293
Lonicera	281
Lophospermum	334
Loranthaceæ	328
Lungwort	313
Lychnis	315
Lycium Vulgare	281
Lyonia	281
Maclura	281
Madagascar Nutmeg	252
Magnolia	251, 282
Acuminata	241
Tripetela	241
Umbrella	241
Mahaleb Cherry	225
Mahonia	282
Maidenhair Tree	297
Mallow Family	329
Malvaceæ	329
Malvaviscus	282
Arboreus	282
Mammilaria	314
Manchurian Aralia	314
Manetti Rose	243
Mangifera	282
Mango	282
Manna Tree	252
Maples, European	233
Japan, Sugar, Etc	250
Red	233
Silver	233
Soft	233
Sycamore	233
Maranta Zebrina	333
Matholia	318
Matrimony Vine	281
Maurandia	334
Mazzard Cherry	225
Meadow Sweet	298
Medlar	227, 283
Melastoma	282, 329
Melastomaceæ	329
Melia Azedarach	282
Melocactus	314
Melon Cactus	314
Mertensia	313
Mesembryantemeæ	329
Mespilus	283
Metamorphosed Flower Stalks	61
Mezereum	269, 283
Microphœnix Saluiti	116
Mignonette	333
Milkwort	272
Mimosa	327
Mimulus	334
Mint	326
Mint Family	326
Missouri Currant	227
Mistletoe, American	328
European	328
Family	328
Mock Orange	285
Monkey Flower	334
Monkshood	332
Monocotyledonous Plants	53
Monœcious Plants	68
Morning Glory	318
Morus	283
Alba	227
Mouse-Ear	316
Mountain Ash	230, 288
Movement and Reorganization of Cells	20
Mulberry	227, 283
Musa	333
Mustard	318
Family	318
Myrabolon Plum	231, 290
Myrtus Communis	283
Myrtle	283
Narcissus	311
Nature of Seeds	73
Nectarine	227
Negundo	283
Aceroides	127
Maple	250
Nelumbium Luteum	330
Speciosum	330
Neottieæ	330
Nerium	283, 312
Nettle Family	334
Nevuisia	185
New Jersey Tea	263
New Zealand Flax	327
Nicotina	334
Nierembergias	334
Nightshade Family	334
Nitrogen	47
Nootka Sound Cypress	264
Norfolk Island Pine	254
Norway Spruce	238
Nymphæa Odorata	330
Nymphæaceæ	329
Nyssa	283
Oak	242, 290
Black	242
Chestnut	242
English	242
European	242
Red	242
White	242
Willow	242
Oak Gall	25
Oak-leaved Hydrangea	276
Oats	323

INDEX. 345

Œnothera 339
Olea 283
    Europæa 283
Oleander 283, 312
Oleaster 270
Olive 227, 283
Onagrarieæ 330
One-celled Alga 13
Onion 327
Ophioglossums 320
Ophrydeæ 330
Opium Poppy 331
Opuntia 314
Orange 226, 265
Orchidaceæ 330
Orchid-Pot 331
Orchis Family 230
Origin and Kinds of Buds 28
Orpine Family 318
Osage Orange 281
Osier 297
Osmanthus aquifolium 284
    Fragrans 284
    Ilicifolius 284
Ostrya 284
    Virginica 284
    Vulgaris 284
Oxalis 322
    Acetosella 32
Oxide of Magnesium 51
Oxydendrum 284
Oxygen 45
Pachysandra 185
Pæony 284, 332
Pæonia Brownii 284
Paliurus 285
Pancratium 311
Pansy 335
Papaveraceæ 321, 331
Papaw 227, 254
Paper Birch 235
Paper Mulberry 258
Papyrus 319
Paradise Stocks 224
Pardanthus 324
Parsley Family 334
Parsnip 334
Passiflora 285
    Edulis 285
    Quadrangularis 285
Passion Flower 285
Patchouly 326
Paulownia Imperialis 285
Peach 228
    Stocks 228
Pecan Nut 260
Pearly Everlasting 317
Pelanquier's Grafter 200
Pelargonium 322

Pemphigus Vitifolia 27
Pentstemon 334
Pepper 334
Pepperidge 265
Pereskia 314
Periwinkle 312
Persimmon 230
Petit's Cleft Grafter 200
Petunia 334
Phaseolus 327
Phillodendron Amurense 285
    Japonicum 285
Phillyrea 286
Philadelphus 285
    Coronarius 286
Phoradendron flavescens 328
Phormium tenax 327
Photinia 286
Phyllocactus 314
Phylloxera Vastatrix 27
Piceas 238
Pinckneya Pubens 286
Pine Apple 314
    Family 314
Pines 237
Pine Tree 286
Pink Family 315
Pinus 286
    Austriaca 237
    Cembra 238
    Coulteri 237
    Excelsa 238
    Densiflora 237
    Flexilis 238
    Mandchurica 238
    Monophylla 57
    Mugho 237
    Ponderosa 237
    Pyrenaica 237
    Resinosa 238
    Sabiniana 57, 237
    Strobus 58, 238
    Sylvestris 237
    Nana 237
Pirus Americana 230, 288
    Angustifolia 224
    Aucuparia 230, 288
    Chinensis 232
    Communis 286
    Japonica 232, 268
    Lusitanica 232
    Malus 224, 286
    Prunifolia 224
    Sinensis 229
Pitcher-Plant Family 333
Pitch Tree 288
Pittosporum Tobira 288
Planera 241, 288
Planer Tree 241, 298

Plane Tree................268
Platanus Occidentalis........288
    Orientalis..............288
Planting the Cutting......152
Pleroma.....................329
Plum....................231, 289
    Stocks..................228
Plumbago....................332
Plumbaginaceæ...............332
Podocarpus Japonica.........288
Poinsettia..............272, 289
Poison Hemlock..............334
Pointed-leaved Willow.......243
Polygamous Plants........... 68
Pomeæ....................... 71
Pomegranate.............232, 290
Poplar......................289
Poppy Family................331
Populus.....................289
Potash...................... 51
Potato......................334
Potentilla..................280
    Fruticosa...............289
Portulacaceæ................332
Portulaca Grandiflora.......332
    Oleracea................332
Portugal Crakeberry.........266
    Quince..................232
Pounder for Firming the Soil..153
Prairie Mallow..............329
Preparing Cuttings..........162
Preservation of Seeds....... 82
Prickly Ash.................308
Pride of India..............282
Prim........................280
Primula.....................332
Primulaceæ..................332
Primrose Family.............332
Privet......................280
Propagating Begonias........167
    House...................156
Propagation by Budding......187
    Different Modes of......135
    by Divisions............176
    by Cuttings.............144
    by Grafting.............199
    by Layers...............170
    by Root-cuttings........180
    by Seeds................136
    by Suckers and Divisions..176
Protoplasm.................. 12
Prunes......................289
Prunus......................231
    Americana...............232
    Domestica...............290
    Chicasa.................232
Pseudolarix.................238
Pseudotsuga.................290
Ptelea Angustifolia.........290

Ptelea Trifoliata...........290
Pterocarya..................290
Pterostyrax.................290
    Hispidum................241
Pulse Family................326
Punica......................290
    Granatum................232
Purple Flowered Raspberry...296
Purposes of Cross Fertilization..129
Pusley, or Purslane.........232
Quaking Aspen...............289
Quamoclit...................318
Queen Plant.................333
Quercus.................242, 290
    Ilex....................242
    Peduculata..............242
    Robur...................242
Quince..................232, 268
    Stocks..................229
Ramanas Rose................334
Ramic Plant.................334
Ranunculaceæ............332, 333
    Asiaticus...............333
Ranunculus Aconitifolius....333
Raphiolepis Japonica........292
Raspberries, Species and Varieties..296
Red Alder...................234
Redbud......................264
Red Cedar...................278
    Maple...................127
    or Norway Pine..........238
    Osier Dogwood...........267
Removing the Bud............193
Resedaceæ...................333
Reseda Odorata..............333
Retinosporas........133, 238, 292
Rhamnus Cathaticus..........292
Rhaphia Ruffia..............189
Rhus Cotinoides.............293
Rhus Cotinus................393
Rhexia......................329
Rhododendron............242, 292
    Catawbiense.............242
    Maximum.................242
    Ponticum................242
Ribes.......................293
    Aureum..................227
    Rotundifolium...........227
Rice........................323
Richardia...................312
Robinia.....................243
    Hispida.................243
    Pseud-acacia........243, 293
Rochea......................318
Rock Cress..................318
    Rose....................275
    Family..................316
Roots and their Functions... 36

Rosa	243, 293
Canina	243
Rugosa	185
Rubiginosa	243
Rose	243, 293
Bay	242, 292
Flowered Raspberry	296
Leaf as a Cutting	33
Mallow	329
of Jericho	16
of Sharon	276
Varieties of	294, 295
Rosemary	326
Roses, Sports Among	134
Round-leaved Gooseberry	227
Rubus, Species and Varieties	296
Rue Family	333
Ruta Graveolens	333
Rutaceæ	333
Rye	323
Sabatier's Implement for Grafting	200
Saddle Grafting	211
Modified	211
Sage	326
Sago Palm	319
Salisburia	297
Salix	243, 297
Acuminata	243
Caprea	243
Purpurea Pendula	243
Salmon Berry	296
Sambucus	297
Sarraceniaceæ	333
Sassafras	280
Officinale	297
Saxifragaceæ	333
Saxifrage Family	333
Scale of Lily Bulb	23
Scarlet Mallow	282
Sciadopitys	297
Scirpus	319
Scitamineæ	333
Scrophulariaceæ	334
Scrub Oak	242
Sea Pink	332
Lavender	332
Sedum	318
Sedge Family	319
Seeds and their Appendages	75
of the Ash	73
of the Elm	74
of Pinus Rigida	77
Size and Vitality	79
Seedling Pine	68
Select List of Plants	249
Selecting Stocks	223
Sempervivums	318
Senna	262
Sensitiveness of Roots	39
Sequoia	297
Gigantea	298
Setting the Cuttings	167
Sex and Fertilization	100
Shadbush	72, 230, 262
Shaddock	226, 265
Shell-bark Hickory	260
Shepherdia Argentea	298
Shrubby Cinquefoil	289
Siberian Pea Tree	235, 259
Saxifrage	333
Side Grafting	221
Silica	50
Silene	315
Silver Bell Tree	275
Leaf	241
Sium Lineare	334
Size and form of Cells	14
Skimmia	298
Slippery Elm	274
Smoke Tree	298
Snapdragon	334
Sneezeworts	317
Snowball Tree	302
Snowberry	299
Snowdrop	311
Tree	240, 275
Soda	51
Solanaceæ	334
Sophora Japonica	243, 298
Sorghum	323
Sour Gum Tree	283
Sop	253
Sorrel Tree	284
Spice Bush	280
Spiderwort Family	316
Spindle Tree	239
Spiræa	298
Aruncus	299
Grandiflora	272
Japonica	333
Opulifolia	299
Splice Grafting the Apple	287
Splice and Tongue Grafting	212
Spoonwood	279
Sports among Roses	134
Spurge Laurel	239, 269
Squill	327
Squirrel Corn	321
Staff Tree	263
Stapelia Glauca	164
Staphylea Bumaldi	299
Pinnata	299
Trifoliata	299
Statice	332
Stems and their Appendages	51
Stick of Buds	189
Sticky Alder	234

Stinking Cedar	301	Thujas	238
St. John's-Wort	276	Thyme	326
St. Julien Plum	231	Tigridia	324
St. Lucie Cherry	225	Tilia	301
Stock or Gilliflower	318	Americana	244
Stocks for Fruit Trees	224	Tillandsia	314
Storax	299	Time to Cut Cions	204
Family	240, 290	Time for Making Layers	174
Stramonium	334	Tobacco	334
Strawberry	71	Tongue or Whip Grafting	212
Flower	71	Torreya Californica	301
Geranium	333	Taxifolia	301
Tree	234, 254	Trabuc's Grafter	200
Strelitza Reginæ	333	Tradescantia	316
Stuartia	299	Virginica	316
Styracaceæ	240, 290	Zebrina	316
Styrax Japonica	240	Transudation of Fluids	16
Officinale	299	Tree of Heaven	152
Sugar Cane	323	Trees, Shrubs and Vines	249
Sumac	293	Tree and Shrub Stocks	232
Sundew Family	319	Tropœolum	322
Sunflower	316	Trumpet Creeper	259, 301
Sweet and Red Bay	280	Trumpet Grape-Gall	27
Sweet Basil	326	Tsuga Canadensis	238
Sweet Briar	243	Tsugas	238
Sweet Chestnut	236	Tulipa	327
Sweet Fern	266	Tulip Tree	281
Sweet Gum Tree	281	Tupelo	283
Sweet Leaf	299	Thunbergia	311
Sweet Potato	318	Turnip	318
Sweet-Scented Shrub	259	Turtlehead	334
Sycamore	288	Ulmus	244, 301
Symphytum	313	Campestris	244
Symphoricarpus	299	Montana	244
Symplocos Tinctoria	299	Umbelliferæ	334
Syringa	285, 299	Umbrella Magnolia	241
Tamarack	280	Pine	297
Tamarindus	300	Urticaceæ	334
Tamarind	300	Vaccinium Macrocarpon	302
Tamarisk	300	Valeriana Edulis	334
Tamarix Gallica	300	Valerinaceæ	334
Tanya	312	Valerian Family	334
Taxodium	300	Vandeæ	330
Distichum	238	Vanillas	331
Taxus Canadensis	300	Variegated Rush	91
Tea Plant	259, 301	Venus' Fly-Trap	320
Tear Tree	269	Veneer Grafting	217
Tecoma	301	Venetian Sumac	293
Radicans	257	Verbenaceæ	335
Terminal and Axillary Buds	30	Vervain Family	335
The Office of Roots	41	Viburnum Opulus	302
Thea	301	Plicatum	302
The Callus	146	Vinca	312
Three-Thorned Acacia	240	Vincent's Grafter	200
Seed	142	Violaceæ	335
Thrift	332	Viola	335
Thorn	267	Odorata	335
Thuja	301	Tricolor	335
Occidentalis	238	Violet Family	335

Virgin's Bower................332	White Thorn as a Stock.........230
Virgilia 265	Wood 230
Viscum Album................328	Wild Black Cherry............226
Vitality of Seeds 77	Red Cherry................226
Experiments to Determine. 81	Olive.......................270
Vitex, Agnus Castus 302	Willow 297
Vitis Vinifera 302	Wind-flower....................332
Wallflower 318	Winged Storax................290
Wahoo 239	Walnut 290
Walnut 241	Witch Hazel...................275
Water Hemlock................334	Wistaria Frutescens............308
Water Lily Family 329	Sinensis 308
Lily 330	Wood Cells...................... 14
Parsnip 334	Sorrel................. 31, 323
Weigela 270, 308	Wormwood 317
Wheat 323	Wych Elm....................244
Whip Grafting................213	Yam Family...................312
White Alder 265	Yankee Budding Knife.........188
Birch 285	Yeast Cells...................... 13
Leaf and Catkin.......256	Yellow Wood..................265
Cedar...............264, 301	Yews 300
Fringe Tree..........287, 265	Zamia Integrifolia.............319
Lily 103	Zanthoxylum Piperitum........308
Mulberry...................227	Zauschnera....................330
Pine 238	Zinnia 316
Thorn......................239